Studies in Computational Intelligence

Volume 500

Series Editor

Janusz Kacprzyk, Warsaw, Poland

For further volumes:
http://www.springer.com/series/7092

Oliver Schütze · Carlos A. Coello Coello
Alexandru-Adrian Tantar · Emilia Tantar
Pascal Bouvry · Pierre Del Moral
Pierrick Legrand
Editors

EVOLVE - A Bridge between Probability, Set Oriented Numerics, and Evolutionary Computation III

 Springer

Editors
Oliver Schütze
Depto. de Computación
CINVESTAV-IPN
Mexico

Carlos A. Coello Coello
Depto. de Computación
CINVESTAV-IPN
Mexico

Alexandru-Adrian Tantar
Computer Science and Communications
    Research Unit
University of Luxembourg
Luxembourg

Emilia Tantar
Luxembourg Centre for Systems
    Biomedicine
University of Luxembourg
Belval
Luxembourg

Pascal Bouvry
Faculty of Sciences, Technology and
    Communication Computer Science and
    Communication Group
University of Luxembourg
Luxembourg

Pierre Del Moral
Bordeaux Mathematical Institute
Université Bordeaux I
Talence cedex
France

Pierrick Legrand
UFR Sciences et Modélisation
Université Bordeaux Segalen
Bordeaux
France

ISSN 1860-949X   ISSN 1860-9503 (electronic)
ISBN 978-3-319-03363-1   ISBN 978-3-319-01460-9 (eBook)
DOI 10.1007/978-3-319-01460-9
Springer Cham Heidelberg New York Dordrecht London

Printed on acid-free paper

Springer is part of Springer Science+Business Media (www.springer.com)

Preface

This volume contains a selection of extended works presented at the international conference EVOLVE, held in August 2012 in Mexico City, México. The aim of the EVOLVE is to build a bridge between probability, set oriented numerics, and evolutionary computation as to identify new common and challenging research aspects.

The conference is also intended to foster a growing interest for robust and efficient methods with a sound theoretical background. EVOLVE aims to unify theory-inspired methods and cutting-edge techniques ensuring performance guarantee factors. By gathering researchers with different backgrounds, a unified view and vocabulary can emerge where the theoretical advancements may echo in different domains.

Summarizing, the EVOLVE conference focuses on challenging aspects arising at the passage from theory to new paradigms and aims to provide a unified view while raising questions related to reliability, performance guarantees, and modeling.

For convenience of the reader, the book is divided into three parts: Probabilistic Modeling (Part I), Evolutionary Computation for Vision, Graphics, and Robotics (Part II), and Multi-objective Optimization (Part III).

Part I contains four contributions. In Chapter 1, Valdez et al. introduce the Empirical Selection Distribution (ESD) for biasing the search of an estimation of distribution algorithm (EDA) based on a Bayesian Network. The authors show that the search can be enhanced by using the empirical selection distribution instead of the standard selection method. The results suggest that the ESD provides more useful information to the algorithm than the usual selection step. Ponce-de-Leon-Senti and Diaz-Diaz compare in Chapter 3 two particular EDA type algorithms with respect to their performance: One that is equipped with a Metropolis step in the inner loop and another one without. Numerical results on four optimization benchmark functions show that both methods are capable of finding the global optimum in practically all cases, but the algorithm with the Metropolis step achieves this goal using less function evaluations.

In Chapter 3, Schaberreiter et al. present a tool implementing a previously proposed Bayesian network based critical infrastructure (CI) risk model which attempts to address the challenges of interdependent CI risk monitoring. The scope of this tool is to provide visual guidance for domain experts to generate a CI risk model from real-world CIs and to simulate/emulate risk scenarios based on this model.

In the last contribution of Part I, Ding and Bouvry survey in Chapter 4 the state-of-the-art in the field of probabilistic modeling for evolving networks. Further, they identify new challenges which emerge on the probabilistic models and optimization strategies in the potential application areas of network performance, network management, and network security for evolving networks.

In the first contribution of Part II (Chapter 5), Olague et al. describe a new approach to synthesize an artificial visual cortex based on what they call brain programming. To be more precise, the authors describe a system composed of an artificial dorsal pathway and an artificial ventral pathway that are fused to create a kind of artificial visual cortex. Experimental results indicate that high recognition rates can be achieved for a well-known multiclass object recognition problem.

In Chapter 6, Hernández et al. present an automatic process for synthesizing visual behaviors by means of genetic programming resulting in specialized prominent point detection algorithms to estimate the trajectory of a camera with a simultaneous localization and map building system. The authors experimentally show that it is in fact possible to find conspicuous points in an image through a visual attention process, and that it is also possible to purposefully generate them through an evolutionary algorithm, seeking to solve a specific task.

In the last contribution of Part II, Olague et al. address in Chapter 7 the problem of evolving an artificial dorsal stream (ADS) using the brain programming strategy. In this work, visual attention is explained as a single mechanism that adapts itself according to a given task, and thus, brain programming is used to design an ADS. As one result, the authors present a solution to the size and missing pop-out problems that were unsolved so far.

Part III consists of four contributions. In the first one, Emmerich and Deutz investigate in Chapter 8 the gradient field of the hypervolume indicator which is the most widely used performance indicator in the field of evolutionary multi-objective algorithms (EMOAs). Their results have a direct impact on local search mechanisms as well as on stopping criteria of hypervolume based EMOAs. Masi and Vasile propose in Chapter 9 an algorithm inspired by the slime mould *Physarum Polycephalum* that is able to incrementally grow decision graphs in multiple directions for discrete multi-objective optimization problems. The resulting algorithm is tested on multi-objective Traveling Salesman and Vehicle Routing Problems. Simulations indicate that building decision sequences in two directions and adding a matching ability (multi-directional approach) is an advantageous choice if compared with the choice of building decision sequences in only one direction (unidirectional approach).

Olofintoye et al. present in Chapter 10 a combined Pareto multi-objective differential evolution algorithm. This algorithm, CPMDE, combines methods of Pareto ranking and Pareto dominance selections to implement a novel selection scheme at each generation. Numerical results show that CPMDE is competitive in solving unconstrained, constrained, and real-world optimization problems.

Finally, in Chapter 11 Salomon et al. further investigate the ability of the recently developed Part and Select Algorithm (PSA) to enhance evolutionary multi-objective algorithms (EMOAs). As main result, the authors show that such a hybridization substantially reduces the risk of the algorithm to fail in finding the Pareto front.

We would like to express our gratitude to the authors who have submitted a contribution to this book. Finally, we would like to thank all the reviewers whose expertly evaluations have helped to maintain the quality of the book.

Mexico City, Luxembourg, and Bordeaux,   *Oliver Schütze*
May 2013   *Carlos A. Coello Coello*
                                             Alexandru-Adrian Tantar
                                                      Emilia Tantar
                                                      Pascal Bouvry
                                                   Pierre Del Moral
                                                   Pierrick Legrand

Organization

Conference General Chairs

Oliver Schütze	CINVESTAV-IPN, México
Alexandru-Adrian Tantar	University of Luxembourg, Luxembourg
Emilia Tantar	University of Luxembourg, Luxembourg
Pascal Bouvry	University of Luxembourg, Luxembourg
Pierre del Moral	INRIA Bordeaux-Sud Ouest, France
Pierrick Legrand	University of Bordeaux 2, France

Advisory Board

Enrique Alba	University of Málaga, Spain
François Caron	INRIA Bordeaux Sud-Ouest, France
Frédéric Cérou	INRIA Rennes Bretagne Atlantique, France
Carlos A. Coello Coello	CINVESTAV-IPN, México
Michael Dellnitz	University of Paderborn, Germany
Frédéric Guinand	University of Le Havre, France
Arnaud Guyader	Université Rennes 2, INRIA Rennes Bretagne Atlantique, France
Arturo Hernandez	CIMAT, México
Günter Rudolph	TU Dortmund University, Germany
Marc Schoenauer	INRIA Saclay - Ile-de-France, University Paris Sud, France
Franciszek Seredynski	Polish Academy of Sciences, Warsaw, Poland
El-Ghazali Talbi	Polytech' Lille, University of Lille 1, France
Marco Tomassini	University of Lausanne, Switzerland
Massimiliano Vasile	University of Strathclyde, Scotland

Programm Committee

Nicola Beume	TU Dortmund University, Germany
Peter A.N. Bosman	CWI, The Netherlands
Jair Cervantes	UAEM-Texcoco, México
Edgar Chavez	University of Michoacana, México
Stephen Chen	York University, Canada
Francisco Chicano	University of Málaga, Spain
David Joaquín Delgado Hernández	Universidad Autónoma del Estado de México, México
Christian Dominguez Medina	CIC-IPN, México
Liliana Cucu-Grosjean	Loria, France
Bernabe Dorronsoro	University of Luxembourg, Luxembourg
Michael Emmerich	Leiden University, The Netherlands
Edgar Galvan	Trinity College University, Ireland
Jesus Gonzalez Bernal	National Institute of Astrophysics, Optics and Electronics, México
Jeffrey Horn	Northern Michigan University, USA
Didier Keymeulen	Jet Propulsion Laboratory, USA
Joanna Kolodziej	University of Bielsko-Biala, Poland
Ricardo Landa Becerra	CINVESTAV-IPN, México
Adriana Lara	IPN, México
Andrew Lewis	Griffith University, Australia
Lili Liu	Northeastern University, China
Francisco Luna	University of Málaga, Spain
Gabriel Luque	University of Málaga, Spain
Evelyne Lutton	INRIA Saclay Ile-de-France, France
Luis Martí	Universidad Carlos III de Madrid, Spain
Jörn Mehnen	Cranfield University, UK
Nicolas Monmarché	University of Tours, France
James Montgomery	Swinburne University of Technology, Australia
Irene Moser	Swinburne University of Technology, Australia
Boris Naujoks	TU Dortmund University, Germany
Sergio Nesmachnow	Universidad de la República, Uruguay
Gustavo Olague	CICESE Research Center, México
Eduardo Rodriguez-Tello	CINVESTAV-IPN, México
Gustavo Sanchez	Simon Bolivar University, Venezuela
Christoph Schommer	University of Luxembourg, Luxembourg
Antonio del Sol	University of Luxembourg, Luxembourg

Juan Humberto Sossa Azuela	CIC-IPN, México
Kiyoshi Tanaka	Shinshu University, Japan
Gregorio Toscano-Pulido	CINVESTAV-IPN, México
Heike Trautmann	TU Dortmund University, Germany
Leonardo Trujillo	CICESE Research Center, México
Alan Reynolds	Heriot-Watt University, Edinburgh, Scotland
Hiroyuki Sato	Shinshu University, Japan
Ponnuthurai Suganthan	Nanyang Technological University, Singapore
Simon Wessing	TU Dortmund University, Germany
Fatos Xhafa	Universitat Politecnica de Catalunya, Spain

Local Organizing Committee

Oliver Schütze	CINVESTAV-IPN, México
Felipa Rosas López	CINVESTAV-IPN, México
Sofía Reza Cruz	CINVESTAV-IPN, México
Erika Berenice Ríos Hernández	CINVESTAV-IPN, México
Santiago Domínguez Domínguez	CINVESTAV-IPN, México
José Luis Flores Garcilazo	CINVESTAV-IPN, México
Christian Dominguez Medina	CIC-IPN, México
Adriana Martinez	Kreaprom S. A. de C. V.

Conference Logo

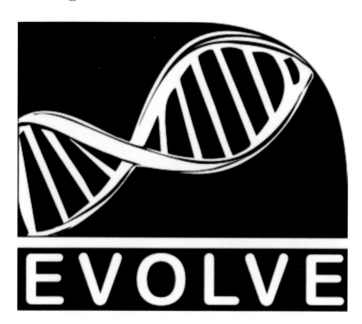

Contents

List of Contributors

Josiah Adeyemo
Department of Civil Engineering and Surveying, Durban University of Technology, Durban, South Africa

Gideon Avigad
ORT Braude College of Engineering, Snunit 51, Karmiel 21982, Israel.

Salvador Botello
Centro de Investigación en Matemáticas A.C., C.P. 36240, Guanajuato, Guanajuato, México

Pascal Bouvry
Interdisciplinary Center for Security, Reliability and Trust, University of Luxembourg, L-1359 Luxembourg, Luxembourg

Eddie Clemente
Departamento de Ciencias de la Computación, División de Física Aplicada, Centro de Investigación Científica y de Educación Superior de Ensenada, Carretera Ensenada-Tijuana No. 3918, Zona Playitas, Ensenada, 22860, B.C., México Tecnológico de Estudios Superiores de Ecatepec, Avenida Tecnológico S/N, Esq. Av. Carlos Hank González, Valle de Anáhuac, Ecatepec de Morelos, México

André Deutz
Leiden University, Leiden Institute for Advanced Computer Science, 2333 CA Leiden, The Netherlands CA Leiden, The Netherlands

Elva Diaz-Diaz
Autonomous University of Aguascalientes, Ave Universidad 940, Colonia Ciudad Universitaria, CP 20131, Aguascalientes, Aguascalientes, México

Jianguo Ding
Interdisciplinary Center for Security, Reliability and Trust, University of Luxembourg, L-1359 Luxembourg, Luxembourg

Christian Domínguez-Medina
Computer Research Center, CIC-IPN, Av. Juan de Dios Bátiz Esq. Miguel Othon de Mendizabal, 07738, Mexico City, México

León Dozal
Departamento de Ciencias de la Computación, División de Física Aplicada, Centro de Investigación Científica y de Educación Superior de Ensenada, Carretera Ensenada-Tijuana No. 3918, Zona Playitas, Ensenada, 22860, B.C., México

Michael Emmerich
Leiden University, Leiden Institute for Advanced Computer Science, 2333 CA Leiden, The Netherlands CA Leiden, The Netherlands

Alan Freitas
Programa de Pós-Graduação em Engenharia Elétrica - Universidade Federal de Minas Gerais - Av. Antônio Carlos 6627, 31270-901, Belo Horizonte, MG, Brasil

Alex Goldvard
ORT Braude College of Engineering, Snunit 51, Karmiel 21982, Israel

Arturo Hernández
Centro de Investigación en Matemáticas A.C., C.P. 36240, Guanajuato, Guanajuato, México

Daniel E. Hernández
Departamento de Ciencias de la Computación, División de Física Aplicada, Centro de Investigación Científica y de Educación Superior de Ensenada, Carretera Ensenada-Tijuana No. 3918, Zona Playitas, Ensenada, 22860, B.C., México

Djamel Khadraoui
Centre de Recherche Public Henri Tudor; Service Science & Innovation (SSI); 29, avenue John F. Kennedy, L-1855 Luxembourg, Luxembourg

Luca Masi
Department of Mechanical & Aerospace Engineering, University of Strathclyde, 75 Montrose Street, G1 1XJ, Glasgow, UK

Arturo Ocampo
Facultad de Estudios Superiores Aragón, UNAM. Av. Rancho Seco s/n. Col. Impulsora, Nezahualcoyotl. Edo. Mex., México

Gustavo Olague
Departamento de Ciencias de la Computación, División de Física Aplicada, Centro de Investigación Científica y de Educación Superior de Ensenada, Carretera Ensenada-Tijuana No. 3918, Zona Playitas, Ensenada, 22860, B.C., México

Oluwatosin Olofintoye
Department of Civil Engineering and Surveying, Durban University of Technology, Durban, South Africa

Fred Otieno
Department of Civil Engineering and Surveying, Durban University of Technology, Durban, South Africa

Eunice Esther Ponce-de-Leon-Senti
Autonomous University of Aguascalientes, Ave Universidad 940, Colonia Ciudad Universitaria, CP 20131, Aguascalientes, Aguascalientes, México

Juha Röning
University of Oulu; Department of Electrical and Information Engineering; P.O.Box 4500, FIN-90014 University of Oulu, Finland

Shaul Salomon
Department of Automatic Control and Systems Engineering, University of Sheffield, Mappin Street, Sheffield S1 3JD, UK.

Thomas Schaberreiter
Centre de Recherche Public Henri Tudor; Service Science & Innovation (SSI); 29, avenue John F. Kennedy, L-1855 Luxembourg, Luxembourg
University of Luxembourg; Computer Science and Communications Research Unit; 6, rue Richard Coudenhove-Kalergi, L-1359 Luxembourg, Luxembourg
University of Oulu; Department of Electrical and Information Engineering; P.O.Box 4500, FIN-90014 University of Oulu, Finland

Oliver Schütze
Computer Science Department, CINVESTAV-IPN, Av. IPN 2508, Col. San Pedro Zacatenco, Mexico City, México

Heike Trautmann
Information Systems and Statistics, University of Münster, Leonardo-Campus 3,
48149 Münster, Germany

S. Ivvan Valdez
Centro de Investigación en Matemáticas A.C., C.P. 36240,
Guanajuato, Guanajuato, México

Massimiliano Vasile
Department of Mechanical & Aerospace Engineering, University of Strathclyde,
75 Montrose Street, G1 1XJ, Glasgow, UK

Part I
Probabilistic Modeling

Effective Structure Learning
in Bayesian Network Based EDAs

S. Ivvan Valdez, Arturo Hernández, and Salvador Botello

Centro de Investigación en Matemáticas A.C.
C.P. 36240, Guanajuato, Guanajuato, Mex.
{ivvan,artha,botello}@cimat.mx

Abstract. Estimation of Distribution Algorithms (EDAs) is a high impact area in evolutionary computation and global optimization. One of the main EDAs strengths is the explicit codification of variable dependencies. The search engine is a joint probability distribution (the search distribution), which is usually computed by fitting the best solutions in the current population. Even though using the best known solutions for biasing the search is a common rule in evolutionary computation, it is worth to notice that most evolutionary algorithms (EAs) derive the new population directly from the selected set, while EDAs do not. Hence, a different bias can be introduced for EDAs. In this article we introduce the so called Empirical Selection Distribution for biasing the search of an EDA based on a Bayesian Network. Bayesian networks based EDAs had shown impressive results for solving deceptive problems, by estimating the adequate structure (dependencies) and parameters (conditional probabilities) needed to tackle the optimum. In this work we show that a Bayesian Network based EDA (BN-EDA) can be enhanced by using the empirical selection distribution instead of the standard selection method. We introduce weighted estimators for the K2 metric which is capable of detecting better the variable correlations than the original BN-EDA, in addition, we introduce formulas to compute the conditional probabilities (local probability distributions). By providing evidence and performing statistical comparisons, we show that the enhanced version: 1) detects more true variable correlations, 2) has a greater probability of finding the optimum, and 3) requires less number of evaluations and/or population size than the original BN-EDA to reach the optimum. Our results suggest that the Empirical Selection Distribution provides to the algorithm more useful information than the usual selection step.

Keywords: Estimation of Distribution Algorithms, Selection Methods Selection Distribution, Empirical Selection Distribution.

1 Introduction

Estimation of Distribution Algorithms (EDAs) are a family of global optimization algorithms which main strengths are related with: the explicit codification of variable dependencies and the use of self-learned parameters for performing

O. Schütze et al. (eds.), *EVOLVE - A Bridge between Probability, Set Oriented Numerics,*
and Evolutionary Computation III, Studies in Computational Intelligence 500,
DOI: 10.1007/978-3-319-01460-9_1, © Springer International Publishing Switzerland 2014

the optimum search. EDAs were derived from probabilistic modeling of genetic algorithms.

In evolutionary computation literature it is well known that the simple genetic algorithm (GA), and most of the standard GA approaches, suffer negative effects of the building block disruption when the decision variables are highly correlated. That is to say, several specific variable instances must be generated or preserved in order to increase the probability of sampling the optimum. The problem was called: the learning linkage problem in the context of evolutionary algorithms [5], and is one of the main motivations of evolutionary computation researchers to propose new methods and frameworks such as Estimation of Distribution Algorithms (EDAs) [14] [17] [24].

Even though the firsts EDAs approaches consider independent variables [14] [6] [2], soon after graphical models were used to improve the search by integrating information about variable correlations [3] [18]. bivariate EDAs shown that they can efficiently solve problems that the simple genetic algorithm and univariate EDAs can not. Regarding the encouraging results, researchers then propose to use more complex models than bivariate, resulting in more powerful algorithms. One of the first algorithms intended to exploit high order correlations was the Factorized Distribution Algorithm [1]. The FDA uses a factorization of the search distribution to perform the search, the original approach does not propose a method to infer such factorization or distribution structure, then other works extended the original to integrate structure-learning methods[21].

One of these approaches is the Bayesian Optimization Algorithm (BN-EDA) [17,11,12]. It is a powerful EDA to solve deceptive-like problems, and in general decomposable or nearly decomposable problems. A general Bayesian Network based algorithm (BN-EDA) is presented in Algorithm 1, [11]. In Step 3 the selection step could use different selection methods to select a subset of the most promising solutions of the current population. In Step 4 the structure and parameters of a Bayesian network are learned from the selected set. The enhancement of these two steps is the main contribution of this article.

Algorithm 1. Bayesian Network based estimation of Distribution Algorithm

---

1 Create a random population $\mathbb{X}^t$ of n_{pop} individuals;
2 Evaluate the population $\mathbb{X}^t$, $\mathbb{F} \leftarrow f(\mathbb{X}^t)$;
3 Select n_{sel} individuals from $\mathbb{X}^t$ using a selection procedure
   $\mathbb{S} \leftarrow selection(\mathbb{X}^t, \mathbb{F})$;
4 Model $\mathbb{S}$ by learning the most adequate Bayesian network B;
5 Create a new population $\mathbb{X}^{t+1}$ by sampling from the joint probability distribution of B;
6 Evaluate population $\mathbb{X}^{t+1}$;
7 Replace all (or some) individuals in population $\mathbb{X}^t$ by those from $\mathbb{X}^{t+1}$;
8 If stopping criteria are not satisfied, return to step 3.

Besides the ability of solving hard optimization problems which require the use of variable dependencies, the BN-EDA provides of additional interesting features, such as:

- It intends to discover an adequate structure of the problem, in order to use it for sampling new high-quality candidate solutions. Considering that the knowledge of the problem structure can be as valuable as the optimum approximation [11], this is a remarkable feature of the BN-EDA.
- There exist model-efficiency techniques which take advantage of the explicit structure codification which leads to important speed ups. For instance, using previous models obtained from several runs to tackle a similar optimization problem [7]. Or, an evaluation relaxation technique [13], which uses an entropy based measure to decide if a candidate solution must be evaluated.
- A priori knowledge could be integrated in the structure and parameter learning procedure.
- For a theoretical Bayesian network based Algorithm called the Factorized Distribution Algorithm (FDA) [16], it has been shown that linear scaling of the population size with n_{var} (number of variables) is sufficient to converge to the optimum, even for hard optimization problems [15].

Nevertheless our article is focus in provide evidence about the enhancements of the empirical selection distribution in the original BN-EDA, the reader must notice that all the improvements and features just listed are shared or can be applied in the original BN-EDA as well as in our approach.

In order to introduce the Empirical Selection Distribution, recall that, convergence to the optimum has been proved for theoretical EDAs [24], by using the exact **selection distribution**.

The selection distribution is defined as the underlying distribution of an infinite sized selected set. It has been shown that the selection distribution depends on the objective function [24]. Thus, the greater the objective function of a point is, the greater the probability associated to such point have to be, in consequence, for the next generation the most fitted individuals are intensively sampled. The selection distribution is defined for infinite sized populations and it can not be directly used in practical approaches.

The **empirical selection distribution** [23] is derived from the theoretical model of the exact selection distribution, but considering a finite sized population. It can be considered as an *a priori* probability or relative frequency for each individual in the population, and then, can be used to estimate the search distribution parameters. Hence, it unifies the selection and estimation steps resulting in a new procedure for biasing the search.

This article introduces a BN-EDA algorithm which uses the empirical selection distribution for estimating the structure as well as the conditional probabilities. According to Mühlenbein [15], a BN-EDA algorithm converges to the optimum if sufficient variable correlations are learned, and there is a large enough probability of sampling the optimum. In this vein, we show, by using the most widely reported objective functions in BN-EDA (by instance: Deceptive-3 and Trap-5), that the approach introduced in this article finds more true variable dependencies than the original BN-EDA, and the probability of finding the optimum is also greater than the original. Additionally, our results show that our approach requires a smaller population size than the original BN-EDA, which impacts on the computational cost for solving hard optimization problems. All the promissory results in this article can be explained as an effect of the empirical selection distribution, considering that it is the unique extra feature we add to the original BN-EDA.

The paper is presented as follows: Section 2 briefly reviews the most common selection methods used in EDAs. Section 3 describes the procedure to integrate the empirical selection distribution for computing the structure and parameters of a BN-EDA EDA. A set of experiments to contrast the empirical selection distribution based BN-EDA(ESD-BN-EDA) with the original BN-EDA is presented in Section 4. Finally, Section 5 presents the main conclusions.

2   Selection Methods

The main goal of a selection operator is to bias the population towards promising regions of the search space. The most common selection methods: truncation, Boltzmann, proportional and tournament are a kind of *subset selection* methods. This kind of method selects a fraction of the population, then some individuals are represented and some others are not. Subset selection can be seen as a weighting method (for the parameter computation) which associates a weight proportional to the number of times an individual is selected, and 0 otherwise. In the case of the truncation method the weights become binary. In the case of the other selection methods, they use a random process to sample the selected set from the population, it is possible (with an small probability, but possible) that even the best solution is not represented in the selected set. Hence, practical approaches which use the subset selection does not maintain a direct relationship between the objective function and representation intensity in the parameter computation, and by consequence neither between the objective function and the posterior search distribution. The possible undesired effects of this lost are:

- Solutions with a high objective value could not be represented or could be misrepresented in the search distribution, because they could not be selected, or their frequencies in the selected set does not correspond with the objective value.
- The selected set very often covers a smaller region than the population, or represents less instances (in discrete space) than the population. Thus a natural variance reduction is expected due to the subset selection [22].
- Information about promising regions could be lost even if there are solutions in the population that indicates such regions, due to the random selection process in most of the subset selection methods.

– Due to the fact that different selected sets could be obtained, by the subset selection method, if it is applied several times over the same population, the performance of the algorithm is not so confident, thus we expect more variance in the performance than methods which always compute the same search distribution from the same population.

On the other hand, the theoretical selection distributions directly depend on the objective function, by instance the proportional selection distribution can be written as Equation 1.

$$p(x_i) = \frac{f(x_i)}{\sum_{j=1}^{n_{instances}} f(x_j)} \tag{1}$$

Where x_i is a possible instance of the decision variables, $f(x_i)$ is the objective function value of x_i, and $n_{instances}$ is the number of possible instances. Note that the exact selection distribution requires to know the objective values of all possible instances.

The empirical selection distribution (ESD) [23] is the counterpart model of the exact selection distribution, when considering a finite population. It intends to explicitly relate the objective function value with the search distribution.

The ESD for the most common selection methods is shown in Table 1. S_t is the selected set, and $p(x_i)$ is a probability associated with the individual i. The ESD has been used in a continuous EDA with a Gaussian distribution [23]. In this article we introduce a Bayesian network (BN) based EDA which uses the ESD. The $p(x_i)$ are used to estimate the structure and parameters of the BN as it is shown in the next section. There are some remarkable features of the ESD:

– It considers the whole population to be computed. Consequently we are using all the available information in the population (notice that subset selection throws away an important fraction of the population). Meaning that all the points in the population are represented.
– Considering that all the population is used, hence, a wider region is covered than using a subset of the population, in consequence we expect a wider variance if needed (if high-fitness individuals are spread over the search space), or a reduced variance (if the high-fitness points are in the same reduced region).
– The sampling intensity correspond to the fitness value.

The next section shows how to integrate the ESD in the structure and parameter computation of a Bayesian network.

3   Estimating the Structure and Parameters of the Bayesian Network

A Bayesian network is a probability model which encode the joint probability distribution for a set of variables. A Bayesian Network for variables

Table 1. Empirical selection distribution model for a finite-sized population

Empirical Selection Distribution
Truncation
$\hat{p}^S(x_i, t) = \begin{cases} \frac{1}{\lvert S_t \rvert} & \text{if } f(x_i) \geq \theta_t \\ 0 & \text{otherwise} \end{cases}$
$\lvert S_t \rvert = \sharp$ of individuals with $f(x_i) > \theta_t$
Proportional
$\hat{p}^S(x_i, t) = \frac{f(x_i)}{\sum_{j=1}^{\lvert X \rvert} f(x_j)}$
Binary Tournament
$\hat{p}^S(x_i, t) = \frac{\sum_{j=1}^{\lvert X \rvert} I(i,j)}{\sum_{i=1}^{\lvert X \rvert} \sum_{j=1}^{\lvert X \rvert} I(i,j)}$
Where $I(i,j) = 1$ if $f(x_j) < f(x_i)$ and 0 otherwise

$\mathbf{X} = \{X_1, ..., X_n\}$ consist of: 1) a network structure $\mathbf{S}$ that encodes a set of conditional independence assertions about variables in $\mathbf{X}$, and 2) a set $\mathbf{P}$ of local probability distributions associated with each variable. The network structure $\mathbf{S}$ is a directed acyclic graph. The nodes in $\mathbf{S}$ are in one-to-one correspondence with the variables $\mathbf{X}$ [9]. Given structure $\mathbf{S}$, the joint probability distribution for $\mathbf{X}$ is given by Equation 2

$$p(x) = \prod_{i=1}^{n} p(x_i \mid \pi_i) \tag{2}$$

The local probability distributions are the corresponding to the terms in the product of Equation 2. Π_i are the parents of the variable X_i.

In order to estimate or learn a Bayesian network from data it is necessary two components: a scoring metric, and a search procedure. The search procedure proposes a candidate Bayesian network, while the scoring metric discriminate among the proposed networks [10].

For this article we use the same search procedure than the original BOA [17], a greedy algorithm with edge addition. The procedure starts with an empty network (without edges), then we test adding a single edge to each variable, the edge that increases the most the scoring metric is actually added to the network. The process is repeated until the metric value can not be increased or the network can not be maintained acyclic.

The scoring metric is modified according to the Empirical Selection Distribution as is shown in the next subsection.

3.1   The K2 Scoring Metric

According to Lima et al. [11] the K2 metric delivers better results for the BOA, in terms of model accuracy, than the BIC metric. Hence, in this article as well as in the original BOA [17] we use the K2 metric for our Bayesian network based algorithm. The K2 metric can be derived from the BDe metric [10] in Equation 3.

$$P(B,S) = P(B) \prod_{i=1}^{n} \prod_{j=1}^{q_i} \left(\frac{\Gamma(N'_{ij})}{\Gamma(N_{ij}+N'_{ij})} \prod_{k=1}^{r_i} \frac{\Gamma(N_{ijk}+N'_{ijk})}{\Gamma(N'_{ijk})} \right) \quad (3)$$

Where: r_i are the number of states of the finite random variable X_i. $q_i = \prod_{X_j \in \Pi_i} r_j$ is the number of possible configurations of the parent set Π_i of X_i. w_{ij} is the $j-th$ configuration of the parents Π_i, $(1 \leq j \leq q_i)$. N_{ijk} is the number of instances in the data S (in this case the selected set), where the variable X_i takes its $k-th$ value x_{ik} and the variables in Π_i take their $j-th$ configuration w_{ij}. $N_{ij} = \sum_{j=1}^{r_i} N_{ijk}$ is the number of instances in the data S where the variables in Π_i take their $j-th$ configuration w_{ij}.

The K2 variant of the BDe metric considers no prior knowledge about the instances. Consequently, $N'_{ijk} = 1$, and $N'_{ij} = r_i$, in Equation 4.

$$P(B,S) = P(B) \prod_{i=1}^{n} \prod_{j=1}^{q_i} \left(\frac{r_i}{\Gamma(N_{ij}+r_i)} \prod_{k=1}^{r_i} \frac{\Gamma(N_{ijk}+1)}{\Gamma(1)} \right) \quad (4)$$

Notice that a logarithmic version of the K2 metric can be use in order to reduce computational errors and effort.

3.2 Modifying the K2 Metric with the ESD

In order to use the ESD for computing the K2 scoring metric, we define a *virtual sample size* greater than n_{pop}, $n_{virtual} >> n_{pop}$. The computation of N_{ijk} and N_{ij} is as shown in Algorithm 2. We show both computation algorithms in the same loop, although it can be in a different loop for practical purposes. Notice that the computation is performed by using a floating point value with $\hat{N}_{ij}$ and $\hat{N}_{ijk}$, and rounded at the end of the computation, in order to reduce rounding errors. Using N_{ij} and N_{ijk} computed as shown in Algorithm 2 we can proceed to compute the K2 metric as usual. $p(x_i)$ in Algorithm 2 is the empirical selection distribution value associated with the individual x_i, for $i = 1..n_{pop}$.

Algorithm 2. Computation of N_{ijk} and N_{ij} using the ESD

---
1 $\hat{N}_{ij} = 0$;
2 $\hat{N}_{ijk} = 0$;
3 **for** *Each individual* $\mathbf{x}$ *in the population* $\mathbb{X}$ **do**
4   **if** $\mathbf{x}_i$ *takes the value* x_{ik} *and the parents* Π_i *takes the value* w_{ij} **then**
5 | $\hat{N}_{ijk} = \hat{N}_{ijk} + (n_{virtual})p(x_i)$;
6   **end**
7   **if** *The parents* Π_i *takes the value* w_{ij} *in* $\mathbf{x}$ **then**
8 | $\hat{N}_{ij} = \hat{N}_{ij} + (n_{virtual})p(x_i)$;
9   **end**
10 **end**
11 $N_{ij} = integer(\hat{N}_{ij} + 0.5)$;
12 $N_{ijk} = integer(\hat{N}_{ijk} + 0.5)$;

3.3   Computing the Conditional Probabilities

Once we have obtained the Bayesian Network structure by using the search process and the scoring metric, then we proceed to compute the parameters or probabilities to sample by using Equation 5.

$$\hat{p}_{ijk} = \hat{P}(X_i = x_{ik}|\Pi_i = w_{ij}) = N_{ijk}/N_{ij} \tag{5}$$

In the case of the ESD-BN-EDA, N_{ij} and N_{ijk} are computed as shown in Algorithm 2. Due to rounding errors, it is possible that a normalization procedure be needed in order to ensure that the parameter is actually a probability (that it sums 1). Then we apply the Equation 6, the sum is for each instance x_{ijk} used to compute the probabilities associated with the variable X_i.

$$p_{ijk} = \frac{\hat{p}_{ijk}}{\sum_{x_{ijk}} \hat{p}_{ijk}} \tag{6}$$

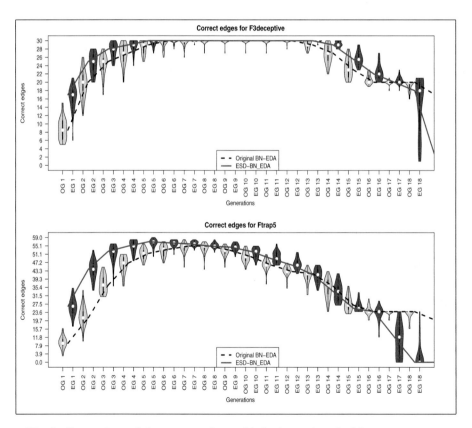

Fig. 1. Comparison of the correct edges added when using the binary tournament

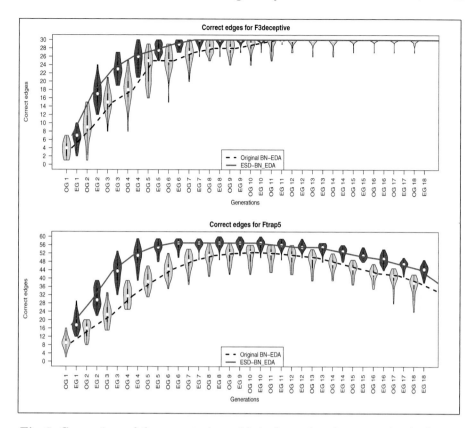

Fig. 2. Comparison of the correct edges added when using the proportional selection

4   Experiments and Performance Analysis

The experiments in this Section are intended to provide evidence about the boosted performance of the BN-EDA when it uses the Empirical Selection Distribution. The information we desire to determine from the experiments is related with three main topics:

1) If the ESD is helpful to determine better the truth structure of the problem. 2) If the ESD increases the probability of finding the optimum, and 3) If the ESD has a beneficial impact on the number of evaluations and population size needed to find the optimum.

The comparisons are performed by using the problems in the original BOA [17], as well as in other research paper which investigate the BOA performance under different selection conditions [11] [12]. The test problems are defined in Table 4

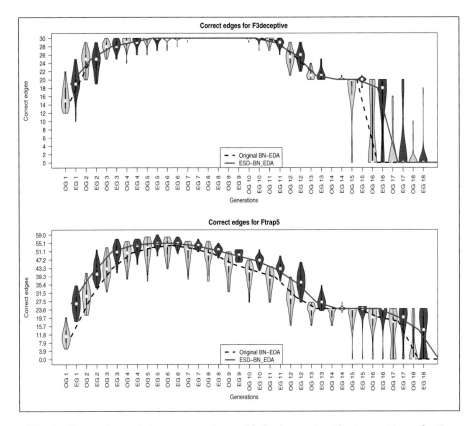

Fig. 3. Comparison of the correct edges added when using the truncation selection

4.1   Evidencing the Number of Correct Dependencies Captured

Experiment Description. For the objective functions presented in Table 4 we define a random order I_i for the variables x_i. This order is used to define the correlations. For example, suppose that we define the order $I = [3, 9, 4, 6, 10, 1, 8, 5, 2, 7]$ for the f_{trap5}, then, the objective function is called using $f_{trap5}(x_3, x_9, x_4, x_6, x_{10})$ and $f_{trap5}(x_1, x_8, x_5, x_2, x_7)$. Hence we expect to add edges in BN which relate the variables in each one of the sets. Under this order we known that the related variables are $[3,9,4,6,10]$ and $[1,8,5,2,7]$. We count how many edges among related variables are added each generation (correct edges). We contrast the correct edges added by the enhanced ESD-BN-EDA versus the original. Using violin plots we graphically show the differences in variance, median, and density (or empirical distribution of the edges) during the generations. Additionally, we perform hypothesis tests to known if the difference between correct edges of ESD-BN-EDA and the original are statistically significant.

Additively composed function: $f(x) = \sum_{i=0}^{l-1} f_k(u)$. where $u = \sum_{j \in S_i} x_j$, and S_i is a partition of 3 or 5 elements of the set $\{1..n_{var}\}$, n_{var} is the number of decision variables, and f_k is one of the following:		
Deceptive order 3	$f_{3deceptive}$	$\begin{cases} 0.9 & if\ u = 0 \\ 0.8 & if\ u = 1 \\ 0 & if\ u = 2 \\ 1 & if\ otherwise \end{cases}$
Trap order 5	f_{trap5}	$\begin{cases} 4 - u & if\ u < 5 \\ 5 & otherwise \end{cases}$

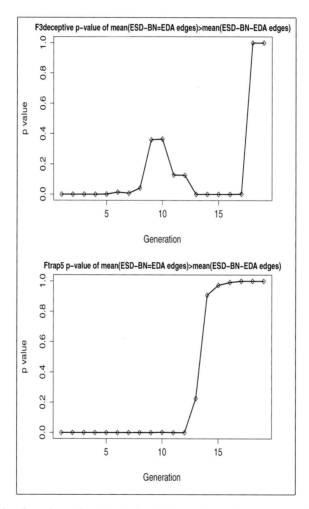

Fig. 4. $p-value$ from hypothesis test about the number of correct edges for the binary tournament

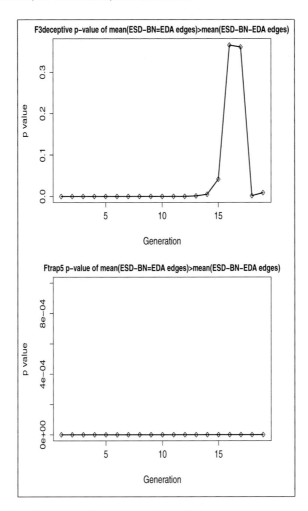

Fig. 5. $p - value$ from hypothesis test about the number of correct edges for the proportional selection

Experiment Settings. We perform 30 independent runs for the test problems in Table 4. The number of variables is 30. We use the first 20 generations for the comparison because this number is enough for convergence of the algorithm. According to [11], the firsts generations is when most of the correct edges are detected, when the convergence is found adding an edge poorly increases the score, and the model is over-fitted adding spurious edges. The population sizes are $\{900, 1300\}$ and the maximum number of allowed parents is $k = \{2, 4\}$ as suggested by Pelikan et al. [17], for the $f_{3deceptive}$ and the f_{trap5} respectively.

Experimental Results. In Figures 1, 2 and 3, we show violin plots of the *correct edges* discovered by each of the algorithms. The violin plots are similar to box plots but they give a better look of the data distribution: the central

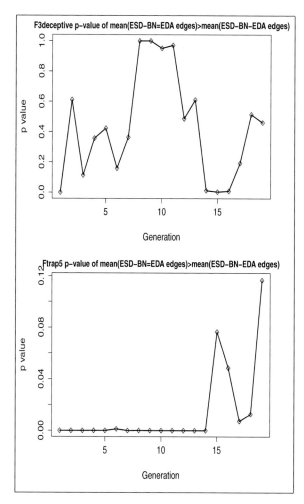

Fig. 6. $p - value$ from hypothesis test about the number of correct edges for the truncation selection

dot is the median of the data, the length of the violin plot measured along the $y - axis$ shows the data dispersion, while the shape of the "box" is a kernel-density approximation to the data density. Hence, the largest plots means the highest variance of the data, in this case the largest violin plots means highest variance in the number of correct edges discovered each generation. The violin plots labeled as "OG x" are computed with data from the x generation of the original BN-EDA, while the violin plots labeled with "EG x" is the corresponding x generation of the Empirical Selection based BN-EDA (ESD-BN-EDA). In Figures 1 to 3, the higher violin plots (with the central dot higher in the $y - axis$) represent the generations and algorithm which discovers more edges. In order to compare more precisely which algorithm discovers more edges, we draw a line

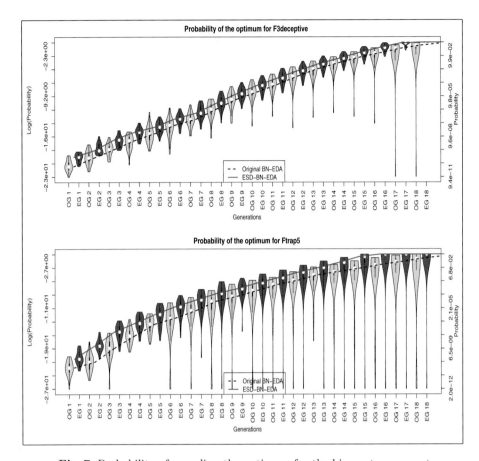

Fig. 7. Probability of sampling the optimum for the binary tournament

with the median of the correct edges discovered, with the same coordinates in the $x - axis$, and the corresponding median in the $y - axis$: the dashed line is the original BN-EDA while the solid line corresponds to the ESD-BN-EDA. Additionally, violin plots let us know which algorithm is the most robust, because more mass around the center as well as a smaller violin, indicate a more consistent performance. For this comparison, another interesting characteristic is that in most of the cases the violin plots show a single mode, which means that most of the times the number of discovered edges is similar for the same generation.

Special Note. When using truncation selection, we perform almost the same estimation for the original BN-EDA and the ESD-BN-EDA, the only difference is the virtual sample and the general procedure of the estimation (for example the calculus of log functions with greater numbers), but the same input information is used to learn the BN. So, we expect a similar behavior of both algorithms

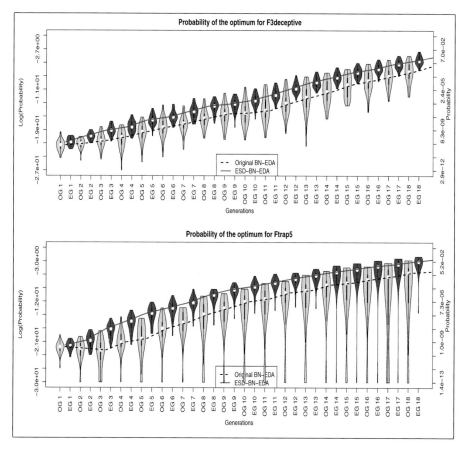

Fig. 8. Probability of sampling the optimum for the proportional selection

for this selection method, and as expected we obtain similar result as they are shown in Figure 3.

The Figures 4, 5 and 6 show the *p-value* from hypothesis tests which compare the means of the number of correct edges discovered each generation by the original BN-EDA versus the ESD-BN-EDA. The hypothesis test is performed by using the Boostrap methodology [4] with 20000 re-samples per generation. In Figure 4 we show the p-value for the mean of correct edges using binary tournament selection. When the p-value is approximately 0, it means that there is strong evidence to say that the ESD-BN-EDA captures more correct edges than the original BN-EDA. Take into account that a 1 value in the p-value does not means that the original BN-EDA captures more correct edges than our approach. A 1 p-value indicates that there is not strong evidence to say the ESD-BN-EDA captures more correct edges. According to Figure 1, we can observe that the ESD-BN-EDA *always* captures at least the same number of correct edges than the original BN-EDA. The 1 p-value in Figure 4 correspond

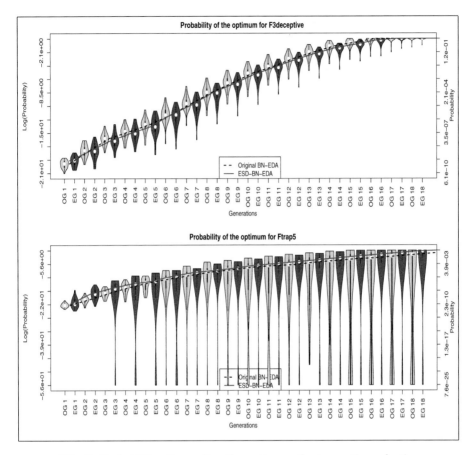

Fig. 9. Probability of sampling the optimum the truncation selection

to the last generations of the algorithms, when the learning of correct edges is not needed because the algorithm has converged.

Figure 4 also shows that the more complex the problem is ($f_{3deceptive} = 3$ correlated variables, $f_{trap5} = 5$ correlated variables), the greater is the evidence to say that the ESD-BN-EDA captures more correct edges than the original.

Figures 5 shows that for proportional selection there is a lot of strong evidence to say that the ESD-BN-EDA captures more correlations. And finally, 6 shows that for truncation selection we can not say which algorithm is better (a lot of $p-values >> 0$), hence as expected, for the truncation selection both algorithms perform quite similar.

4.2   Evidencing the Probability of Finding the Optimum

Recall that the goal of these experiments is not to show the effectiveness of the BN-EDA, which has been largely tested [17,8,20,19], in contrast we intend to show that the BN-EDA can be enhanced by using the ESD. An obvious

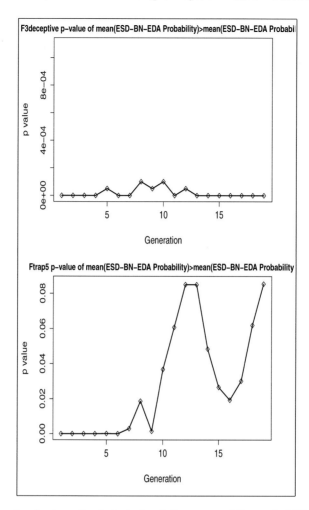

Fig. 10. $p - value$ for hypothesis test about the mean of the probability of sampling the optimum for the binary tournament

enhancement is to increase the probability of finding the optimum. Each generation is a different learning stage, because a cumulative bias is introduced in the model via the selection method, because of this, we test the probability of finding the optimum each generation in order to show a consistent increment of it when using the ESD. Using the same settings than in the experiment above, we compute for each generation the probability of finding the optimum. The findings are reported in Figures 7, 8 and 9. These Figures show the violin plots for the probability of the optimum, according to the computed structure and conditional probabilities of the Bayesian network, using the original BN-EDA and the ESD-BN-EDA, for 30 independent runs. In addition, the median of this probability is plotted by using a solid line for the ESD-BN-EDA and a dashed

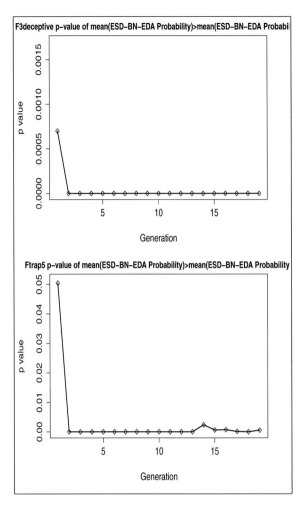

Fig. 11. $p - value$ for hypothesis test about the mean of the probability of sampling the optimum for the proportional selection

line for the original, the x coordinate is the same for both algorithms thus the lines reflect the actual difference between the medians. As can be seen there is evidence to say that the ESD-BN-EDA has a higher probability of finding the optimum during the whole generations. In order to show that the statistical evidence is strong enough we perform Boostrap hypothesis tests. The Figures 10, 11 and 12 are the p-value of testing that the probability of the optimum using the ESD is greater than the original counterpart.

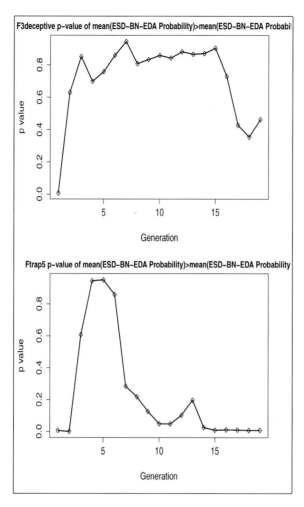

Fig. 12. $p - value$ for hypothesis test about the mean of the probability of sampling the optimum for the truncation selection

A Numerical Issue of This Test. Suppose that, according to the BN discovered, x_i depends on x_j at generation t. Assume the frequencies of the instances as follows: $freq(x_i = 0, x_j = 0) = 50$, $freq(x_i = 0, x_j = 1) = 50$, $freq(x_i = 1, x_j = 0) = 100$ and $freq(x_i = 1, x_j = 1) = 0$. Using the Bayes rule $p(x_i = 1|x_j = 1)$ is not defined, and the empirical joint probability of $p(x_i = 1, x_j = 1) = 0$. But notice that the empirical marginal probability, $p(x_i = 1) = 100/200 = 0.5$ and $p(x_j = 1) = 50/200 = 0.25$. Thus if the same frequencies (or quite similar) are maintained for the next generation, and x_i does not depend on x_j (given the new computed structure), then $p(x_i = 1, x_j = 1) = p(x_i)p(x_j) = (0.5)(0.25) = 0.125 > 0$. In practical terms this means that in a

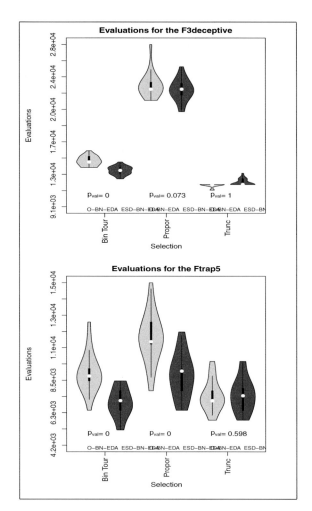

Fig. 13. $p - value$ and violin plots for comparison of the number of evaluations, for different selection methods. The hypothesis tested is mean(evaluations of BN-EDA)>mean(evaluations of ESD-BN-EDA).

generation the probability of sampling the optimum (under a given structure) could be 0, and some generations later the probability could be different from 0. Thus, for practical comparisons (because the violin plots of the probability are in log scale) we replace the probabilities equal to 0, with the worst value found in the same run, this only affects the plots (not the hypothesis test, neither the conclusions of the experiment) by avoiding to graphically report $-\infty$ values.

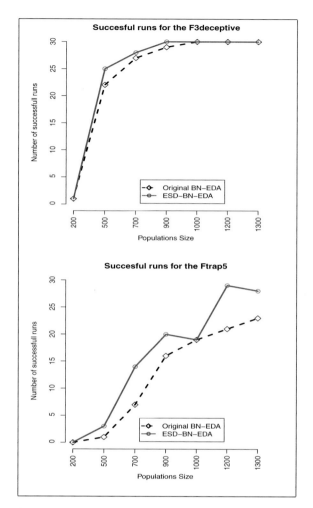

Fig. 14. Succesful runs from 30, for different population sizes using the binary tournament selection

Results of the Experiment Which Test the Probability of Finding the Optimum. As can be seen in Figures 7 to 12 there is sufficient statistical evidence to say that the ESD-BN-EDA has a greater probability to find the optimum than the original. The explanation is that the most fitted solutions actually are sharing variable values with the optimum and they have a greater weight to compute the Bayesian network structure and the conditional probabilities. The p-value close to 0 in the first generations is an remarkable indicator, because this stage of the algorithm is crucial for detecting the interest region where the optimum is. In this stage the EDA is performing a more intense exploration than in the last generations in which the algorithm basically is refining the optimum approximation and converging to a stable point. As expected the hypothesis

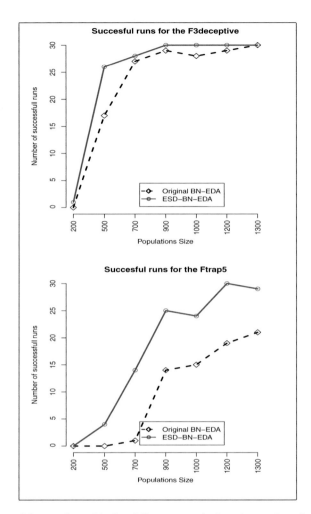

Fig. 15. Succesful runs from 30, for different population sizes using the proportional selection

test for the truncation method are not conclusive, because as mentioned for this selection the ESD-BN-EDA and the original are using the same information.

4.3   Evidencing the Reduction of the Number of Function Evaluations

In this experiment we run the BN-EDA and the ESD-BN-EDA, under the same parameters than the experiments above, but the algorithm was stopped when it finds the optimum. Then we present similar violin plots than in the experiments above, for the number of evaluations needed for each of the algorithms to reach the optimum. We use the same settings than the experiment above, except the

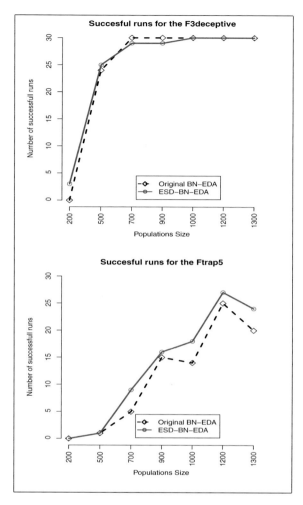

Fig. 16. Succesful runs from 30, for different population sizes using the truncation selection

population size which is fixed in 1300. 20 successful runs were used for the violin plots and the hypothesis tests. We tested if the mean of the number of evaluations of the original BN-EDA is greater than ESD-BN-EDA counterpart.

Results of the Experiment Which Test the Number of Evaluations. As can be seen the number of evaluations is less for the ESD-BN-EDA than the original one. And it is sufficient statistical evidence to support this conclusion. Additionally, the number of evaluations are not statistically different for the truncation method.

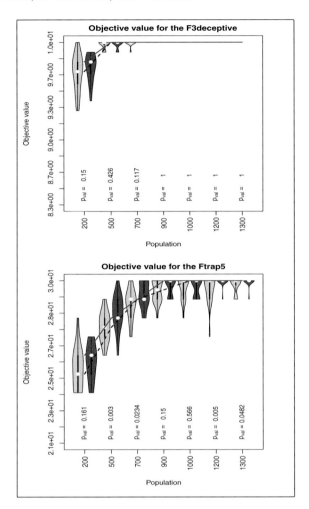

Fig. 17. Objective value for different population sizes for the binary tournament selection. The p-value correspond to the hypothesis test mean(f_{best} of ESD-BN-EDA) >mean(f_{best} of BN-EDA).

4.4   Evidencing the Reduction of the Population Size

In this experiment we use different population sizes and observe two different results:

1. The number of times the optimum is reached with each population.
2. The best objective value (in distribution via violin plots) found with each population. Also we perform hypothesis test to know if there is sufficient statistical evidence to say that the one algorithm delivers a better objective value.

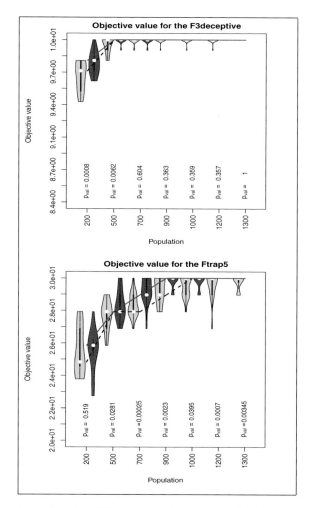

Fig. 18. Objective value for different population sizes for the proportional selection. The p-value correspond to the hypothesis test mean(f_{best} of ESD-BN-EDA) >mean(f_{best} of BN-EDA).

Results of the Experiment Which Test the Population Sizes. Figures 14, 15 and 16 show the number of times the optimum is reached with different population sizes, contrasting the original BN-EDA with ESD-BN-EDA. As can be see, the ESD-BN-EDA consistently outperform the BN-EDA. Figures 17, 18 and 19, show the objective function values of the elite individual for both approaches. Even if we do not always have sufficient statiscal evidence according to the p-value, it can be seen that many times the performance of the ESD-BN-EDA is the best, and in other cases it is a least as good as the original approach.

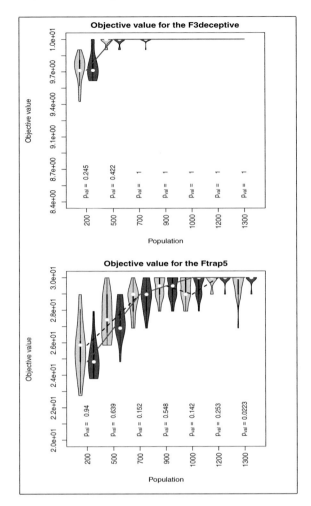

Fig. 19. Objective value for different population sizes for the truncation selection. The p-value correspond to the hypothesis test mean(f_{best} of ESD-BN-EDA) >mean(f_{best} of BN-EDA).

5   Conclusions

In this paper we deeply test the Empirical Selection Distribution (ESD) integrated in the selection-estimation step of the Bayesian Network based EDA(BN-EDA). According to our experiments the ESD significantly boosts the BN-EDA. The main conclusions are when using the Binary tournament and the proportional selection we collect sufficient statistical evidence to say that the ESD improves the BN-EDA performance. Several enhancements have been proposed for the BN-EDA [11,7,13], in this context the ESD enhances the BN-EDA, and at the same time, it is allowed to be combined with the other enhancements.

The ESD can be seen as a general selection method for biasing EDAs, in this article we show that even for complex models such as Bayesian Networks, integrating the ESD in an EDA can be relatively simple. Additionally, we test the ESD with three selection methods but notice that in general the ESD are weights for each individual. Then, other kind of bias such as Pareto raking for multi-objective approaches, or diversity measures can be used, in order to solve multiobjective problems, or to increase the impact of diverse solutions in the parameter estimation. In summary, the ESD enhances the BN-EDA but is not restricted to be used in it. Additionally, the ESD is not only a enhancement for an EDA but a way of inducing the bias in the search, hence researchers could propose different biasing schemes only by modifying the ESD computation, and maintaining unaltered the main EDA body. The results obtained by the boosting of the BN-EDA with the ESD, encourage the future work in using it with other multivariate discrete and continuous EDAs. Future work considers to integrate non-standard selection methods in the BN-EDA, for instance, methods which consider diversity measures. Additionally, we will continue exploring the effects of the ESD in other EDAs.

References

1. The Factorized Distribution Algorithm for additively decomposed functions, vol. 1 (1999)
2. Baluja, S.: Population-based incremental learning. Tech. Rep. CMU-CS-94-163, Computer Science Department, Carnegie Mellon University, Pittsburgh, PA (June 1994)
3. Bonet, J.S.D., Isbell Jr., C.L., Viola, P.A.: MIMIC: Finding optima by estimating probability densities. In: Advances in Neural Information Processing Systems 9, NIPS, pp. 424–430. MIT Press (1996)
4. Efron, B.: The Jacknife, the Bootstrap and Other Resampling Plans. Society for Industrial and Applied Mathematics, 1400 Architect's Building, 117 South 17th Street, Philadelphia, Pensilvania (1982)
5. Harik, G., Goldberg, D.E.: Learning linkage. In: Proceedings of the 4th Workshop on Foundations of Genetic Algorithms, pp. 247–262 (1996)
6. Harik, G.R., Lobo, F.G., Goldberg, D.E.: The compact genetic algorithm. IEEE Trans. Evolutionary Computation 3(4), 287–297 (1999)
7. Hauschild, M.W., Pelikan, M., Sastry, K., Goldberg, D.E.: Using previous models to bias structural learning in the hierarchical boa. Evolutionary Computation 20(1), 135–160 (2012)
8. Hauschild, M., Pelikan, M., Sastry, K., Lima, C.: Analyzing probabilistic models in hierarchical boa. Trans. Evol. Comp. 13(6), 1199–1217 (2009)
9. Heckerman, D.: A tutorial on learning with Bayesian networks. Tech. Rep. MSR-TR-95-06, Microsoft Research, Advanced Technology Division, Microsoft Corporation (1995)
10. Heckerman, D., Geiger, D., Chickering, D.M.: Learning Bayesian Networks: The Combination of Knowledge and Statistical Data. Machine Learning 20(3), 197–243 (1995)
11. Lima, C., Lobo, F., Pelikan, M., Goldberg, D.: Model accuracy in the bayesian optimization algorithm. Soft Computing - A Fusion of Foundations, Methodologies and Applications 15, 1351–1371 (2011)

12. Lima, C.F., Pelikan, M., Goldberg, D.E., Lobo, F.G., Sastry, K., Hauschild, M.: Influence of selection and replacement strategies on linkage learning in boa. In: IEEE Congress on Evolutionary Computation, pp. 1083–1090 (2007)
13. Luong, H.N., Nguyen, H.T.T., Ahn, C.W.: Entropy-based efficiency enhancement techniques for evolutionary algorithms. Information Sciences 188, 100–120 (2012)
14. Mühlenbein, H., Paaß, G.: From recombination of genes to the estimation of distributions I. Binary parameters. In: Voigt, H.M., Ebeling, W., Rechenberg, I., Schwefel, H.P. (eds.) PPSN 1996. LNCS, vol. 1141, pp. 178–187. Springer, Heidelberg (1996)
15. Mühlenbein, H.: Convergence theorems of estimation of distribution algorithms. In: Shakya, S., Santana, R. (eds.) Markov Networks in Evolutionary Computation. ALO, vol. 14, pp. 91–108. Springer, Heidelberg (2012)
16. Mühlenbein, H., Mahnig, T.: FDA -a scalable evolutionary algorithm for the optimization of additively decomposed functions. Evolutionary Computation 7(4), 353–376 (1999)
17. Pelikan, M., Goldberg, D.E., Cantú-Paz, E.: BOA: The Bayesian Optimization Algorithm. In: Banzhaf, W., Daida, J., Eiben, A.E., Garzon, M.H., Honavar, V., Jakiela, M., Smith, R.E. (eds.) Proceedings of the Genetic and Evolutionary Computation Conference GECCO 1999, vol. I, pp. 525–532. Morgan Kaufmann Publishers, San Fransisco (1999)
18. Pelikan, M., Mühlenbein, H.: The Bivariate Marginal Distribution Algorithm. In: Advances in Soft Computing – Engineering Design and Manufacturing, pp. 521–535 (1999)
19. Pelikan, M., Sastry, K., Goldberg, D.E.: Scalability of the bayesian optimization algorithm. International Journal of Approximate Reasoning 31(3), 221–258 (2002)
20. Pelikan, M., Sastry, K., Goldberg, D.E.: iBOA: the incremental bayesian optimization algorithm. In: Proceedings of the 10th Annual Conference on Genetic and Evolutionary Computation, GECCO 2008, pp. 455–462. ACM, New York (2008)
21. Santana, R.: A Markov Network Based Factorized Distribution Algorithm for Optimization, pp. 337–348 (2003)
22. Shapiro, J.L.: Diversity loss in general estimation of distribution algorithms. In: Runarsson, T.P., Beyer, H.-G., Burke, E.K., Merelo-Guervós, J.J., Whitley, L.D., Yao, X. (eds.) PPSN 2006. LNCS, vol. 4193, pp. 92–101. Springer, Heidelberg (2006)
23. Valdez-Peña, S.I., Hernández-Aguirre, A., Botello-Rionda, S.: Approximating the search distribution to the selection distribution in EDAs. In: Proceedings of the 11th Annual Conference on Genetic and Evolutionary Computation, GECCO 2009, pp. 461–468. ACM, New York (2009)
24. Zhang, Q., Muhlenbein, H.: On the convergence of a class of estimation of distribution algorithms. Trans. Evol. Comp. 8(2), 127–136 (2004)

Optimization by Structure Learning during Algorithm Execution Using an Adaptive Extended Tree Cliqued – EDA (AETC – EDA)

Eunice Esther Ponce-de-Leon-Senti and Elva Diaz-Diaz

Autonomous University of Aguascalientes, Ave Universidad 940, Colonia Ciudad Universitaria, CP 20131, Aguascalientes, Aguascalientes, Mexico
{eponce,ediazd}@correo.uaa.mx

Abstract. The objective of this chapter is to compare, with respect to the performance two versions of EDA type algorithm (AETC – EDA) whose probabilistic distribution structure is learned at each step using the Extended Tree Adaptive Learning (ETreeAL) Algorithm. One of the versions has a Metropolis step in the inner loop, and the other do not. The samples are generated evaluating the objective function, and running Boltzmann selection roulette through all the cliques of the learned model. In the outer loop the temperature parameter is updated. The efficiency is tested using 4 benchmark functions known by its difficulty for evolutionary algorithms. The experiments were performed with 50 and 100 variables. As results of the experiments, the algorithms obtain the optimum in practically in all the cases, both algorithms use the learned cliques at each step of the inner cycles to obtain the best solution at hand. The algorithm with Metropolis step uses less evaluations in all cases, except for the Fc_2 function.

Keywords: Estimation of Distribution Algorithms, Discrete Graphical Markov Model, Linkage Learning, Structure Learning, Evolutionary Algorithms.

1 Introduction

Most of the studies about EDAs algorithms perform an empirical comparison of two or more algorithms, employing a set of benchmark problems and some performance criteria, for example, the number of evaluations. The theoretical basis (concepts, definitions, and theorems) needed to identify the features that led to a better performance of one algorithm with respect to other for each type of problem, are not frequently analyzed. In this chapter, the theoretical, practical and experimental issues are put together, to offer information for enriching the discussion.

In genetics, linkage is the tendency for alleles of different genes to be passed together from one generation to the next [30]. For the evolutionary algorithms it is interesting to detect linkage groups for the underlying structure of the optimization problem. If the linkage groups are not known in advance they must be detected during the algorithm execution. This question of linkage learning was proposed by Holland [9]. He

O. Schütze et al. (eds.), *EVOLVE - A Bridge between Probability, Set Oriented Numerics, and Evolutionary Computation III,* Studies in Computational Intelligence 500,
DOI: 10.1007/978-3-319-01460-9_2, © Springer International Publishing Switzerland 2014

noted that the complexity of the adaptive systems comes from the interactions of alleles that reflects the adaptation of the genotype to the environment. This adaptation requires most of the time a nonlinear structure that changes with the change of the simultaneous appearance of groups of variable alleles (epistasis - some phenotype appears only with an exact combination of alleles). One of the formal models proposed by Holland to reflect this property is the discrete probabilistic model defined over the space of all the possible combinations of alleles. Specifically a Discrete Graphical Markov Model (DGMM) lends his structure of interactions to represent the linkage. The last decade appeared a lot of optimization algorithms based on the linkage learning using probabilistic graphical models, some of the most used are: the estimation of distribution algorithms (EDAs) [19], the bivariate marginal distribution algorithm (BMDA) [22], the Bayesian optimization algorithm [21], the Estimation of Bayesian Network Algorithm (EBNA) [7], the Learning Factorized Distribution Algorithm (LFDA) [18], and EDAs based on Markov networks and Gibbs sampling [26] [27]. There are other methods to learn the linkage: the perturbation methods are based on perturbing the variables, and examining the fitting differences, to detect sets of variables to be linked [29]. The linkage identification by nonlinearity check detect nonlinearity by pairwise perturbations in order to identify the linkage set [20] [28]. The use of DGMM is based on the hypothesis, that learning the structure of the graphical model is equivalent to linkage learning as will be seen in the next section. The DGMM is a type of hierarchical log linear probabilistic graphical model [14].

This chapter presents an analysis of the roll and performance of the linkage learning in the AETC – EDA optimization algorithm [23]. The AETC – EDA consists on two main parts. The first part is the ETreeAL algorithm used to learn the linkage with a probabilistic graphical model [6], and the second part is a cliqued Gibbs sampler algorithm (CG – Sampler) that uses the cliques of the graphical model learned in the first part to generate the next population. The strategy is to use a DGMM, meaning it, that in the course of optimization the algorithm iteratively obtains the structure of a graphical model (linkage learning) and use this information to generate samples using a simulation algorithm based on an annealing process in the outer cycle and a cliqued Gibbs sampler in the inner cycle. The evaluation of each solution generated by the cliqued Gibbs sampler and the new structure obtained at each generation, guides the search to the global optimum. The cliqued Gibbs sampler uses a Boltzmann selection procedure [8] in order to apply a Boltzmann roulette selection defined in this chapter. Each algorithm is tested with four deceptive functions. The experiments are performed with two versions of the cliqued Gibbs sampler algorithm, one with a Metropolis step and the other one without it.

In Section 2 some definitions and concepts needed to describe the algorithms are given. In Section 3 the algorithm pseudocodes are described and explained. In Section 4, the test functions are described. In Section 5, the experiments are designed, and the parameters explained. In Section 6, the results are presented and discussed, and in Section 7 the conclusions are given.

2   Definitions and Concepts

The fundamental definitions and concepts needed to construct and explain the algorithms are detailed in order to make the chapter self contained. More explanation about the AETC – EDA can be consulted in [6], [23], and [24].

Definition 1. *Let $S = \{s_1, s_2, ..., s_v\}$ be **a set of sites** and let $\mathscr{G} = \{\mathscr{G}_s, s \in S\}$ be the neighborhood system for S, meaning it any collection of subset of S for which*
  1) $s \notin \mathscr{G}_s$ and
  2) $s \in \mathscr{G}_r \Leftrightarrow r \in \mathscr{G}_s$.
*$\mathscr{G}_s$ is **the set of neighbors of** s, and **the pair** $\{S, \mathscr{G}\}$ **is a graph**.*

Definition 2. *A subset $C \subset S$ is **a clique** if every pair of distinct sites in C are neighbors. $\mathscr{C} = \{C\}$ denotes the set of cliques. Let $X = \{x_s, s \in S\}$ denotes any family of random binary variables indexed by S. Let Ω be the set of all possible values of X, that is, $\Omega = \{w = (x_1, x_2, ..., x_v) : x_i \in \{0,1\}\}$ is the **sample space** of all possible realizations of X.*

Definition 3. *X is **a Markov random field (MRF) with respect to** $\mathscr{G}$ if*

$$P(X = w) > 0 \text{ for all } w \in \Omega \tag{1}$$

and

$$P(X_s = x_s | X_r = x_r, r \neq s) = P(X_s = x_s | X_r = x_r, r \in \mathscr{G}_s) \text{ for every } s \in S \text{ and } w \in \Omega \tag{2}$$

where X denotes the random variable and w denotes the values that this variable can take.

Definition 4. *A **Gibbs distribution relative to** $\{S, \mathscr{G}\}$ is a probability measure π on Ω with the following representation*

$$\pi(w) = \frac{1}{Z} e^{-U(w)/T} \tag{3}$$

where Z and T are constants. U is called the energy function and has the form

$$U(w) = \sum_{c \in \mathscr{C}} V_c(w) \tag{4}$$

Each V_c is a function on Ω that only depends on the coordinates x_s of w for which $s \in C$, and $C \subset S$.

Definition 5. *The family $\{V_c, c \in \mathscr{C}\}$ is called **a potential** and the constant*

$$Z = \sum_w e^{-U(w)/T} \tag{5}$$

*is called **the partition function**. The constant T is named temperature and it controls the degree of "peaking" in the density π.*

Theorem 1. *(Equivalence theorem) Let $\mathcal{G}$ be a neighborhood system. Then X is a MRF with respect to $\mathcal{G}$ if and only if $\pi(w) = P(X = w)$ is a Gibbs distribution with respect to $\mathcal{G}$.*

A more extensive treatment can be seen in [8], [3], [10].

Definition 6. *Gibbs Sampling is a Markovian updating scheme that works as follows. Given an arbitrary starting set of values $x^{(0)} = (x_1^{(0)}, x_2^{(0)}, ..., x_v^{(0)}) \in \Omega$, one value $x_i^{(0)}$ is drawn from the conditional distribution $P(x_i | x_1^{(0)}, ..., x_{i-1}^{(0)}, x_{i+1}^{(0)}, ..., x_v^{(0)})$ for each $i = 1, ..., v$, where v is the number of variables. So, each variable is visited in the natural order until v. After that, a new individual $x^{(1)} = (x_1^{(1)}, x_2^{(1)}, ..., x_v^{(1)})$ is obtained.*

Geman and Geman [8] demonstrated that for

$$t \to \infty, \quad x^{(t)} \to x$$

where $x = (x_1, x_2, ..., x_v)$ and t is the parameter of the process (if the process is an algorithm, t is an iteration). This sampling schema required v random variate generations, one for each state i of the schema.

As a component of the cliqued Gibbs sampler, the generator part, a selection procedure is used.

The well-known Metropolis step is a fundamental part of the Simulating Annealing (SA) algorithm. It was first introduced by Metropolis [15] to simulate the physical annealing process of solids.

Definition 7. *The Metropolis step is defined as the decision of accepting a new state X' based on the α criterion defined by:*

$$\alpha = e^{-\delta U(X)/kT} \tag{6}$$

where

$$\delta U(X) = U(X') - U(X) \tag{7}$$

T denotes the temperature and k the Boltzmann constant. For $k = 1$, at each T the SA algorithm aims to draw samples from the Boltzmann equilibrium distribution:

$$\pi_T(X)e^{-\delta U(X)/kT} \tag{8}$$

Definition 8. *Annealing Adaptive Search (AAS): The AAS is an iterative algorithm that simulated a non stationary finite state Markov chain whose state space is the domain of the cost function to be optimized. The iterations depend on the annealing schedule of temperatures $\{T_k\}$ that goes to zero and gives the name of annealing to the algorithm. The process converges to a family of Boltzmann distributions. The algorithm at each iteration generates a random solution according to a Boltzmann distribution depending of the parameter T_k. When T_k goes to zero the generated solutions converges to the optimum [8].*

Definition 9. *Let x_i be a solution and $f(x_i)$ be the function of x_i to optimize (the cost function), then the **Bolztmann selection** procedure evaluates a solution x_i*

$$P(x_i) = \frac{\exp(-f(x_i))}{E(x_m)} \tag{9}$$

where $E(x_m) = \sum_j f(x_j) p(x_j)$ is the expected value of all the elements in the current population.

$P(x_i)$ is not invariant to scaling but is invariant to translation [5].

Given the state i of the schema $x^{(t)}$, for the value $P[x_i | x_1^{(t)}, \ldots, x_{i-1}^{(t)}, x_{i+1}^{(t)}, \ldots, x_v^{(t)}]$ a value for $x_i^{(t+1)}$ is obtained selecting as follows: if $p(x_i^{(t+1)}) > p(x_i^{(t)})$ select $x_i^{(t+1)}$, else calculate

$$q = \frac{p(x_i^{(t+1)})}{p(x_i^{(t)})}, \tag{10}$$

and if a random number uniformly generated in the interval $[0,1]$ is less that q, select $x_i^{(t+1)}$.

Definition 10. *The **Boltzmann roulette selection**, select one individual from a contingency table cell, according to the Boltzmann selection procedure (Definition 9).*

Definition 11. *The **K-L divergence from the probability model M to the data** x is given by the Kullback-Leibler information [12]*

$$G^2(M,x) = \log(L(\widehat{m}_n^M(x))) = -2\sum_{i=1}^{k} x_i \log_2\left(\frac{\widehat{m}_i^M}{x_i}\right). \tag{11}$$

where k is the sample number of different individuals, n is the total number of individuals, and $\widehat{m}_n^M$ is the maximum-likelihood parameter estimator of m_n^M.

The K-L divergence is also known as relative entropy, and can be interpreted as the amount of information in the sample x not explained by the model M, or the deviation from the model M to the data x. This K-L divergence is used to calculate *SMCI* (Definition 13) and in time *EMUBI* (Definition 14) in the next paragraphs.

Definition 12. *The **mutual information measure** $I_{X_iX_j}$ for all $X_i, X_j \in X$ is defined as*

$$I_{X_iX_j} = I(X_i, X_j) = \sum_{x_i, x_j} P(x_i, x_j) \log \frac{P(x_i, x_j)}{P(x_i)P(x_j)}, \tag{12}$$

where $P(x_i, x_j) = P(X_i = x_i, X_j = x_j)$.

As a part of a strategy to learn a graphical Markov model, a statistical model complexity index (*SMCI*) is defined and tested by Diaz et al. [6]. Based on this index it is possible to obtain an evaluation of the sample complexity and to prognose the graphical model to explain the information contained in the sample.

The model representing the uniform distribution is denoted by M_0, and the model represented by a tree is denoted by M_T. If a sample is generated by a model M containing more edges than a tree, the information about the model M contained in this sample and not explained, when the model structure is approximated by a tree, may be assessed by the index defined as follows.

Definition 13. *Let x be a sample generated by a model M. The **statistical model complexity index** (SMCI) of the model M is defined by the quantitative expression [6]*

$$SMCI(M,x \mid M_T) = \frac{G^2(M_T,x) - G^2(M,x)}{G^2(M_0,x)}. \tag{13}$$

This index can be named sample complexity index, because it is assessed by the quantity of information contained in the sample generated by the model M.

Definition 14. *Let G be the graph of the model M, and let v be the number of vertices, let $MNE(v)$ be the maximum number of edges formed with v vertices. The **edge missing upper bound index** (EMUBI) (see [6]) is defined by*

$$EMUBI_\tau(M,x \mid M_T) = \tau(MNE(v) - v + 1)SMCI(M,x \mid M_T), \tag{14}$$

where τ is the window allowing a proportion of variability in the sample to get into the model. This coefficient is a filter that allows the sample relevant information for the model structure construction [6].

This index is used to prognose the number of edges to add to a tree in order to approximate the complexity of the sample using the graphical model, in this case it is an unrestricted graphical Markov model.

The next two definitions are used to assess the similarity of the linkage structure learned by the algorithm and the linkage structure of the optimization problem.

Definition 15. *Given the learned graph G_1, and the generator graph G_2, **a graph similarity index of G_1 respect to G_2**, $GSI(G_1, G_2)$ is given by the number of common edges divided by the number of edges in the generator graph G_2 [6]. Denote by E_1 the set of edges from G_1 and by E_2 the set of edges from G_2, then the similarity index is given by:*

$$GSI(G_1, G_2) = \frac{|E_1 \cap E_2|}{|E_2|}, \tag{15}$$

where $|C|$ denotes the number of elements in the set C.

Definition 16. *Let $\mathcal{L} = \{l_i\}$ be the learned graph, and $\mathcal{G} = \{g_j\}$ the generator graph given by their cliques l_i and g_j respectively*

a. *If for all j there exists i such that $l_i \supset g_j$ then it is said that $\mathcal{L}$ **overlearns** $\mathcal{G}$.*
b. *If for some (but not all) j there exists i such that $l_i \supset g_j$ then it is said that $\mathcal{L}$ **partially learns** $\mathcal{G}$, and*
c. *The proportion of cliques learned is the number of g_j cliques contained in some l_i cliques, divided by $|\mathcal{G}|$, is named **index of learned cliques or learning index**.*

Algorithm 1. Index of learned cliques algorithm

   Input: $\mathscr{L}$ and $\mathscr{G}$

1 Calculate $|\mathscr{L}|$ and $|\mathscr{G}|$

2 count $\leftarrow 0$

3 Learning index $\leftarrow 0$

4 **for** $i = 1$ *to* $|\mathscr{G}|$ **do**

5   **for** $j = 1$ *to* $|\mathscr{L}|$ **do**

6   **if** $g_j \subset l_i$ **then**

7   count $\leftarrow$ count $+1$

8   **end**

9   **end**

10 **end**

   Output: Learning index $\leftarrow \dfrac{\text{count}}{|\mathscr{G}|}$

The generator graph in this chapter is the graph given by the cliques used as variable blocks of the objective function. To calculate the index of learned cliques an algorithm whose pseudocode is given in Algorithm 1, is used.

To calculate and interpret the learning index it is necessary to note that a graphical model is constructed over hypergraphs [2] that are formed by subsets of a given set. Graphical models are a particular type of hypergraphs whose maximal subsets correspond to cliques of a graph. So, it makes sense to compare cliques subsets with variables blocks.

3 The Adaptive Extended Tree Cliqued – EDA (AETC – EDA)

In this section the algorithms that conform the AETC – EDA are explained. The CL algorithm (Algorithm 2) is used as first step of the EtreeAL (Algorithm 3) in order to obtain the tree model. The tree model is extended to an unrestricted graphical model by EtreeAL. After that the CG – sampler (Algorithm 4) uses the graphical model learned to generate the next population of the AETC – EDA (Algorithm 5).

The CL algorithm (Chow and Liu algorithm) [4] obtains the maximum weight spanning tree using the Kruskal algorithm [11] and the mutual information values $I_{X_iX_j}$ (Definition 12) for the random variables. The tree obtained by this algorithm is denoted by $M_T(CL)$.

Algorithm 2. CL algorithm

   Input: Distribution P over the random vector $X = (X_1, X_2, ..., X_v)$

1 Compute marginal distributions P_{X_i}, $P_{X_iX_j}$, for all $X_i, X_j \in X$.

2 Compute mutual information values (Definition 12) $I_{X_iX_j}$ for all $X_i, X_j \in X$.

3 Order the values from high to low (w.r.t.) mutual information.

4 Obtain the maximum weight spanning tree $M_T(CL)$ by the Kruskal algorithm [11].

   Output: The maximum weight spanning tree $M_T(CL)$

The CL-algorithm (see Algorithm 2) calculates for the number of variables v, $\frac{v(v-1)}{2}$ mutual informations. If each variable takes two values then the CL-algorithm has complexity $O(\frac{v(v-1)}{2}2^2) = O(2v(v-1)) = O(v^2)$. The ETreeAL (Algorithm 3) adds the number of prognosed edges to the tree to obtain the model structure necessary to the CG-sampler input, and these operations have polynomial complexity.

Algorithm 3. The extended tree adaptive learning algorithm (ETreeAL)

Input: Distribution P over the random vector $X = (X_1, X_2, ..., X_v)$

1 Call CL Algorithm in order to obtain the maximum spanning tree using the mutual information measure $I_{X_iX_j}$ (Definition 12) as weight edges.

2 Calculate the edge missing upper bound prediction index $(EMUBI_\tau(M, x \mid M_T(CL)))$.

3 Add to $M_T(CL)$, τ percent from missing edges in the order of the mutual information values (Definition 12).

Output: Extended tree model structure $M_{EXT}(CL)$

The Algorithm 3 obtains the Markov model adapted to the sample, and gives the cliques of this model. The Algorithm 4 has 4 parameters: T is the temperature, $\alpha \in (0, 1)$ is the annealing parameter, C is the bound of the iterations number to the annealing step of the temperature, and Q is the upper bound of the outer cycle to complete the population size. This algorithm receives the cliques of the graphical model as input to fulfill the condition of equivalence required by Theorem 1 (see Section 2). The variables of each clique are used together by the CG – sampler optimizer to generate new individuals for the sample. Other authors used blocked Gibbs sampler [25], and that is why the name "Cliqued Gibbs Sampler" was given to this sampler. Let $f(x)$ be the objective function. $\widehat{E}(x_m)$ is the $f(x)$ mean estimator obtained for the population of solutions. Without losing generality in the description of the CG-Sampler algorithm, we assume that the optimization problem is to obtain a minimum and a convenient change can be made to obtain a maximum.

Observing the CG – Sampler Optimizer Algorithm (Algorithm 4) it is seen that the algorithm input besides the cliques, receives a population of selected solutions, let P be the population and let the structure of the graphical model adjusted be given by its cliques, $\mathscr{C} = \{c_1, ..., c_k\}$. In the outer cycle the iterations run through all elements of the population (N). The inner loop runs through the number of cliques (k) and the Step 7 generates a marginal table for each clique, and calculates a Boltzmann roulette selection, running over the marginal table corresponding to the clique c_j, for all j=1,..,k. Let $CM = \max_{c_j \in \mathscr{C}} |c_j|$ then the worst case is when $|c_j| = CM$ for all j. In this case the complexity is given by $O((N)(k)2^{CM})$, so the complexity of the CG-Sampler is exponential in the size of the maximum clique of the model.

3.1 Adaptive Extended Tree Cliqued – EDA (AETC – EDA) Pseudocode

The AETC – EDA (Algorithm 5) employs the ETreeAL to obtain the Markov model structure of the population of solutions for each algorithm's iteration, with this structure the CG- sampler optimizer obtains the next population.

Algorithm 4. CG – Sampler Optimizer Algorithm

Input: A population of selected solutions, and the structure of the graphical model, given by its cliques, and parameters temperature T, α, C, and Q

1 $P_{New} \leftarrow \{\}, i \leftarrow 1, S^+ \leftarrow S_1, j \leftarrow 0, q \leftarrow 0$

2 **repeat**

3   Take the solution S_i of the selected solutions

4      $S\prime \leftarrow S_i$

5   **for** $k = 1$ *to Number of cliques* **do**

6   Take the clique ω_k

7   Generate the marginal table of the clique ω_k:

8   **if** $f(\omega_k(S\prime)) > S^+$ **then**

9              $S^+ \leftarrow \omega_k(S\prime)$

10             $P_{New} \leftarrow P_{New} \cup \{S^+\}, j \leftarrow j+1$

11   **end**

12   Roulette Selection: Select the individual $\omega_k(S\prime)_l$ corresponding to the l cell of the marginal table according to its selection probability (Boltzmann Selection) Definition 9.

13         $S\prime \leftarrow \omega_k(S\prime)_l$

14   **end**

15   /* Metropolis Step */

16   **if** $f(S\prime) > \widehat{E}(x_m)$ **then**

17   **if** $S\prime \notin P_{New}$ **then**

18             $P_{New} \leftarrow P_{New} \cup \{S\prime\}, j \leftarrow j+1$

19   **end**

20   **end**

21   **else**

22   **if** $Random \leq \exp -\frac{f(S_i)-f(S\prime)}{\widehat{E}(x_m)*T}$ **then**

23   **if** $S\prime \notin P_{New}$ **then**

24                 $P_{New} \leftarrow P_{New} \cup \{S\prime\}, j \leftarrow j+1$

25   **end**

26   **end**

27   **else**

28   **if** $S_i \notin P_{New}$ **then**

29                 $P_{New} \leftarrow P_{New} \cup \{S_i\}, j \leftarrow j+1$

30   **end**

31   **end**

32   **end**

33   **if** $i = (Percent\ of\ selection)*N/100$ **then**

34         $i = 0, q \leftarrow q+1$

35   **end**

36   **else**

37         $i \leftarrow i+1$

38   **end**

39   **if** $q \equiv 0 \bmod C$ **then**

40         $T \leftarrow T * \alpha$

41   **end**

42   **if** $q > Q$ **then**

43   *break*

44   **end**

45 **until** *(a new population P_{New} with size N is obtained)*;

Output: A new population P_{New} (Gibbsian Population and S^+)

Algorithm 5. Adaptive Extended Tree Cliqued – EDA (AETC – EDA)

Input: Number of variables, function to optimize, population size, percent of
selected individuals, stop criterion, τ, temperature T and α

1   Create the initial population of size N at random.

2   **repeat**

3   Evaluate the population.

4   Order the population and select a portion.

5   With the selected portion of the population call the *ETreeAL* (Algorithm 3).

6   Call cliqued Gibbs sampler (CG – Sampler) optimizer (Algorithm 4)

7   **until** *(Some stop criterion is met)*;

    Output: Solution of the optimization problem

---

4   Test Functions

Four functions are employed to analyze the performance of the algorithm AETC –
EDA (see Algorithm 5). The selected functions are a sample from functions of different
difficulties. All of them are deceptive functions proposed to study the genetic algorithm
performance. The Overlapping $Trap_5$ is the most difficult one in this analysis.

In all functions we use v as the number of the variables, x_i is a binary variable for
every i, and $u = \sum x_i$ is the number of ones in the solution $\mathbf{x} = (x_1,...,x_v)$.

F_{c_2} Deceptive Problem. Proposed in [17] its auxiliary function and deceptive decomposable function are as follows.

$$f_{Muhl}^5 = \begin{cases} 3.0 \text{ for } x = (0,0,0,0,1) \\ 2.0 \text{ for } x = (0,0,0,1,1) \\ 1.0 \text{ for } x = (0,0,1,1,1) \\ 3.5 \text{ for } x = (1,1,1,1,1) \\ 4.0 \text{ for } x = (0,0,0,0,0) \\ 0.0 \text{ otherwise} \end{cases}$$

$$f_{c_2}(\mathbf{x}) = \sum_{i=1}^{\frac{v}{5}} f_{Muhl}^5(x_{5i-4},x_{5i-3},x_{5i-2},x_{5i-1},x_{5i}) \qquad (16)$$

The optimum solution is the string with all positions put in 0.

F_3 Deceptive Problem. This problem has been proposed in [17]. Its auxiliary function
and deceptive decomposable function are as follows.

$$f_{dec}^3 = \begin{cases} 2 \text{ for } u = 0 \\ 1 \text{ for } u = 1 \\ 0 \text{ for } u = 2 \\ 3 \text{ for } u = 3 \end{cases}$$

$$f_{3deceptive}(\mathbf{x}) = \sum_{i=1}^{\frac{v}{3}} f_{dec}^3(x_{3i-2},x_{3i-1},x_{3i}) \qquad (17)$$

The optimum solution is the string with all positions put in 1.

Trap$_k$ Problem. A Trap function of order k [21] can be defined as

$$Trap_k(u) = \begin{cases} k & u = k \\ k - 1 - u, \, otherwise \end{cases}$$

$$f_{Trap_k}(\mathbf{x}) = \sum_{i=1}^{\frac{v}{k}} Trap_k(x_{5i-4}, x_{5i-3}, x_{5i-2}, x_{5i-1}, x_{5i}) \tag{18}$$

We use $k = 5$. The optimum solution is the string with all positions put in 1.

OverlappingTrap$_5$ Problem. An Overlapping Trap function of order 5 [20] can be defined as

$a_i = Trap_5(x_{5i-4}, x_{5i-3}, x_{5i-2}, x_{5i-1}, x_{5i})$
$b_i = Trap_5(x_{5i+1}, x_{5i+2}, x_{5i+3}, x_{5i+4}, x_{5i+5})$
$a_{v/5} = Trap_5(x_{5(v/5)-4}, x_{5(v/5)-3}, x_{5(v/5)-2}, x_{5(v/5)-1}, x_{5(v/5)})$
$b_0 = Trap_5(x_1, x_2, x_3, x_4, x_5)$

$$f_{OverlappingTrap_5}(\mathbf{x}) = \sum_{i=1}^{\frac{v}{5}-1} [a_i + \omega\phi(a_i + b_i)] + a_{v/5} + \omega\phi(a_{v/5} + b_0) \tag{19}$$

where $\omega = 1$ and ϕ is defined as

$$\phi(a+b) = \begin{cases} -1 \text{ if } (a+b) \equiv 0 \bmod 2 \\ +1 \text{ if } (a+b) \equiv 1 \bmod 2 \end{cases}$$

The optimal solutions are constructed with blocks of size 5. There are two different optimal solutions $\mathbf{x}_1$ and $\mathbf{x}_2$. $\mathbf{x}_1 = $ (000001111100000....0000011111) and $\mathbf{x}_2 = $ (111110000011111....1111100000), where v (number of variables) fulfills the restriction $v \equiv 0 \bmod 5$.

For the F_{c_2} function, the graphical model assumed has 10 cliques, the number of edges is 100, and all cliques have a size 5. For the F_3 Deceptive function, the graphical model assumed has 16 cliques, the number of edges is 48, and the size of the cliques is 3. The $Trap_5$ has the same graphical model as the F_{c_2} function. In the case of $OverlappingTrap_5$ the graphical model is more difficult to describe in terms of number of cliques, size of cliques and number of edges.

5   Experimental Design

Two versions of the algorithm AETC – EDA are tested, with and without Metropolis step. To test the proposed algorithms two experiments are designed using different parameter values and four test problems described in the Section 4. The first experiment employs 48 and 50 variables and population sizes of 50, 70, 90, and 500 individuals, and the second one employs 99 and 100 variables, and populations sizes of 140, 180, and 1000 individuals. For each combination of factors, 30 replications are used. To assess the performance of the algorithms the number of evaluations, the learning index, the similarity index, and the mean best solution value are calculated. In the Table 1 the descriptions of the AETC – EDA parameters are listed, where the learning index and similarity index are given to help the discussion of the results. For the two experiments, the τ parameter of the ETreeAL (Algorithm 3) is fixed to 1, the number of generation is 500, and the selection percent of the population is 60 in AETC – EDA.

Table 1. AETC – EDA parameters description

Parameters	Description
Number of Evaluations	Mean number of evaluation in 30 runs
Number of Cliques	Mean Number of cliques of learned graph in 30 runs
Size of Cliques	Mean size of cliques of learned graph in 30 runs
Edge number	Mean number of edges of learned graph in 30 runs
Learning Index	Mean rate in 30 runs, of cliques of the generator graph contained in some one clique of the learned graph, divided by cliques number of generator graph (See Definition 16)
Similarity Index	Mean Similarity in 30 runs, of the learned graph respect to the generator graph (See Definition 15)
Mean best value	The mean best value obtained by the AETC – EDA in 30 runs
Sample Complexity x	x be the complexity of a Population of AETC – EDA in each generation and run (See Definition 13). The x is bounded by the value after the sign $<$

6   Experimental Results and Discussion

The Tables 2, 3, 4 and 5 contain the results of the algorithm AETC – EDA without a Metropolis step, and using 48 and 50 variables. In the Table 2 the results for 48 and 50 variables with a population size of 50, are given and the optimum is obtained in all cases. In the Table 3 the results for 48 and 50 variables and a population size of 70 are presented. Note that the mean size of cliques used are less with a population of 70 than with a population of 50. In the Table 4 the results for 48 and 50 variables and a population size of 90 are presented and the tendency to use less mean size of cliques is observed. In the Table 5 the results for 48 and 50 variables and a population size of 500 and the tendency to use a size of cliques is fast equal to 2 can be clearly seen.

The Tables 6, 7, 8 and 9 contain the results of the algorithm using 48 and 50 variables and employing a Metropolis step. In the Table 6 the results of 48 and 50 variables with a population size of 50 are given, and the optimum is obtained in all cases. Note that the size of cliques is more than 2, but the mean size of cliques is less than in the case without the Metropolis step. In the Tables 7, 8, and 9 the tendency to decrease the mean size of cliques continue to arrive short to 2.

In the Tables 10, 11, 12 and 13 the number of variables are 99 and 100, and the algorithm do not contain a Metropolis step. In the Table 10 the population size is 100,

Table 2. AETC – EDA (CG - Sampler without a Metropolis step). Population size= 50.

Parameters	F_{c_2}	F_3	$Trap_5$	$Overlapping Trap_5$
Variables	50	48	50	50
Number of Evaluations	364.233	31731.6	86022.033	27268.666
Number of Cliques	37.2	35.9	43.933	60.5
Size of Cliques	2.586	2.869	2.414	3.706
Edge number	81.03	89.733	71.266	175.433
Learning Index	0.464	0.509	0.387	0.171*
Similarity Index	0.573	0.926	0.389	0.585*
The Optimum value	40	48	50	55
Mean best value	40	48	50	55
Sample Complexity x	$x < 0.07$	$x < 0.3$	$x < 0.3$	$x < 0.4$

Table 3. AETC – EDA (CG - Sampler without a Metropolis step). Population size= 70.

Parameters	F_{c_2}	F_3	$Trap_5$	$Overlapping Trap_5$
Variables	50	48	50	50
Number of Evaluations	594.966	42616.033	112614.133	24704.666
Number of Cliques	37.633	33.7	41.066	49.6
Size of Cliques	2.439	2.774	2.385	3.142
Edge number	72.633	79.733	66.433	123.6
Learning Index	0.548	0.582	0.506	0.332*
Similarity Index	0.578	0.972	0.491	0.583*
The Optimum value	40	48	50	55
Mean best value	40	48	50	55
Sample Complexity x	$x < 0.04$	$x < 0.2$	$x < 0.2$	$x < 0.2$

Table 4. AETC – EDA (CG - Sampler without a Metropolis step). Population size= 90.

Parameters	F_{c_2}	F_3	$Trap_5$	$Overlapping Trap_5$
Variables	50	48	50	50
Number of Evaluations	682.366	70882.333	127519.233	25284.3
Number of Cliques	39.866	34.366	38.166	42.433
Size of Cliques	2.342	2.629	2.529	3.03
Edge number	67.066	72.433	74.466	108.533
Learning Index	0.591	0.608	0.553	0.401*
Similarity Index	0.557	0.972	0.588	0.632*
The Optimum value	40	48	50	55
Mean best value	40	48	50	55
Sample Complexity x	$x < 0.03$	$x < 0.09$	$x < 0.2$	$x < 0.2$

Table 5. AETC – EDA (CG - Sampler without a Metropolis step). Population size= 500.

Parameters	F_{c_2}	F_3	$Trap_5$	$Overlapping\,Trap_5$
Variables	50	48	50	50
Number of Evaluations	4341.6	8383.166	4654.266	4456.466
Number of Cliques	48.7	45.2	48.033	47.3
Size of Cliques	2.006	2.04	2.021	2.042
Edge number	49.366	48.80	50.033	51.033
Learning Index	0.676	0.523	0.672	0.668*
Similarity Index	0.403	0.704	0.41	0.42*
The Optimum value	40	48	50	55
Mean best value	40	48	50	55
Sample Complexity x	$x < 0.001$	$x < 0.004$	$x < 0.003$	$x < 0.004$

Table 6. AETC – EDA (CG - Sampler with a Metropolis step). Population size= 50. $T = 8000$. $\alpha = 0.9$.

Parameters	F_{c_2}	F_3	$Trap_5$	$Overlapping\,Trap_5$
Variables	50	48	50	50
Number of Evaluations	278.833	2112.8	7868.533	2879.5
Number of Cliques	43.333	50.8	46.333	59.633
Size of Cliques	2.432	2.594	2.138	2.594
Edge number	76.666	96.2	56.166	107.1
Learning Index	0.338	0.329	0.611	0.245*
Similarity Index	0.436	0.762	0.43	0.443*
The Optimum value	40	48	50	55
Mean best value	40	48	50	55
Sample Complexity x	$x < 0.05$	$x < 0.2$	$x < 0.2$	$x < 0.2$

Table 7. AETC – EDA (CG - Sampler with a Metropolis step). Population size= 70. $T = 8000$. $\alpha = 0.9$.

Parameters	F_{c_2}	F_3	$Trap_5$	$Overlapping\,Trap_5$
Variables	50	48	50	50
Number of Evaluations	364.966	2931.7	9790.733	2627.266
Number of Cliques	41.6	45.133	44.033	48.466
Size of Cliques	2.279	2.516	2.247	2.488
Edge number	64.566	82.233	61.866	87.766
Learning Index	0.502	0.414	0.591	0.34*
Similarity Index	0.478	0.825	0.481	0.489*
The Optimum value	40	48	50	55
Mean best value	40	48	50	55
Sample Complexity x	$x < 0.03$	$x < 0.08$	$x < 0.1$	$x < 0.07$

Table 8. AETC – EDA (CG - Sampler with a Metropolis step). Population size= 90. $T = 8000$. $\alpha = 0.9$.

Parameters	F_{c_2}	F_3	$Traps_5$	$Overlapping Traps_5$
Variables	50	48	50	50
Number of Evaluations	522.533	3825.066	10160	1890.466
Number of Cliques	42.9	40.5	40.566	43.366
Size of Cliques	2.207	2.438	2.35	2.445
Edge number	60.233	70.233	66.933	76.233
Learning Index	0.588	0.485	0.555	0.411*
Similarity Index	0.482	0.829	0.53	0.5*
The Optimum value	40	48	50	55
Mean best value	40	48	50	55
Sample Complexity x	$x < 0.02$	$x < 0.05$	$x < 0.06$	$x < 0.07$

Table 9. AETC – EDA (CG - Sampler with a Metropolis step).Population size= 500. $T = 8000$. $\alpha = 0.9$.

Parameters	F_{c_2}	F_3	$Traps_5$	$Overlapping Traps_5$
Variables	50	48	50	50
Number of Evaluations	6452.566	4228.7	3344.866	3403
Number of Cliques	49	46	48.4	48.166
Size of Cliques	2	2.021	2.012	2.018
Edge number	49	48	49.6	49.966
Learning Index	0.677	0.516	0.674	0.673*
Similarity Index	0.4	0.687	0.406	0.409*
The Optimum value	40	48	50	55
Mean best value	40	48	50	55
Sample Complexity x	$x < 0.0007$	$x < 0.002$	$x < 0.002$	$x < 0.002$

in the Table 11 the population size is 140, in the Table 12 the population size is 180, and in the Table 13 is 1000. The tendency to decrease the size of cliques needed by the algorithm continues, and when the size of the population is 1000, the mean size of cliques arrives short to 2.

The tables 14, 15, 16 and 17 the number of variables is 99 and 100 and the Metropolis step is contained in the algorithm. The tendency to use less size of cliques with growing population sizes continues to arrive short to 2 with population size of 1000.

The similarity index in the first experiment using 48 and 50 variables is less in the case of the algorithm with Metropolis step than without Metropolis step. This can be seen comparing the Tables 2, 3, 4 and 5 with the Tables 6, 7, 8 and 9. The same behavior can be seen in the second experiment using 99 and 100 variables (see the Tables 10, 11, 12 and 13 and the Tables 14, 15, 16 and 17). On the other hand, the learning index, in both experiments is increased, when the population size is increased too.

Table 10. AETC – EDA (CG - Sampler without a Metropolis step). Population size= 100.

Parameters	F_{c_2}	F_3	$Trap_5$	$Overlapping\,Trap_5$
Variables	100	99	100	100
Number of Evaluations	3411.1	324764	625310.666	325852.8
Number of Cliques	100.066	81.833	90.7	116.6
Size of Cliques	2.713	2.319	2.15	3.23
Edge number	216.533	128.533	118.366	267.066
Learning Index	0.244	0.461	0.301	0.236*
Similarity Index	0.494	0.722	0.285	0.438*
The Optimum value	80	99	100	110
Mean best value	80	99	100	110
Sample Complexity x	$x < 0.05$	$x < 0.2$	$x < 0.3$	$x < 0.1$

Table 11. AETC – EDA (CG - Sampler without a Metropolis step). Population size= 140.

Parameters	F_{c_2}	F_3	$Trap_5$	$Overlapping\,Trap_5$
Variables	100	99	100	100
Number of Evaluations	3819.6	722050.6	848906.533	401630.266
Number of Cliques	82.033	85.433	84.633	87.133
Size of Cliques	2.497	2.181	2.232	2.762
Edge number	165.266	114.6	128.8	191.066
Learning Index	0.403	0.508	0.37	0.392*
Similarity Index	0.528	0.743	0.396	0.543*
The Optimum value	80	99	100	110
Mean best value	80	99	100	110
Sample Complexity x	$x < 0.04$	$x < 0.06$	$x < 0.2$	$x < 0.2$

Table 12. AETC – EDA (CG - Sampler without a Metropolis step). Population size= 180.

Parameters	F_{c_2}	F_3	$Trap_5$	$Overlapping\,Trap_5$
Variables	100	99	100	100
Number of Evaluations	7461.566	1450325.6	945505.133	484636.8
Number of Cliques	78.166	89.7	81.3	85.533
Size of Cliques	2.419	2.116	2.312	2.667
Edge number	149.233	109.533	135.7	179.9
Learning Index	0.483	0.499	0.459	0.467
Similarity Index	0.552	0.709	0.491	0.589*
The Optimum value	80	99	100	110*
Mean best value	80	99	100	110
Sample Complexity x	$x < 0.02$	$x < 0.06$	$x < 0.07$	$x < 0.09$

Table 13. AETC – EDA (CG - Sampler without a Metropolis step). Population size= 1000.

Parameters	F_{c_2}	F_3	$Trap_5$	$Overlapping\,Trap_5$
Variables	100	99	100	100
Number of Evaluations	152299.233	1874572.8	26963.8	111745.8
Number of Cliques	94.9	90.566	95.166	93.6
Size of Cliques	2.052	2.082	2.05	2.094
Edge number	104.866	105.433	104.266	108.8
Learning Index	0.66	0.534	0.661	0.651*
Similarity Index	0.429	0.741	0.426	0.445*
The Optimum value	80	99	100	110
Mean best value	80	99	100	110
Sample Complexity x	$x < 0.002$	$x < 0.004$	$x < 0.002$	$x < 0.004$

Table 14. AETC – EDA (CG - Sampler with a Metropolis step). Population size= 100. $T = 8000$. $\alpha = 0.9$

Parameters	F_{c_2}	F_3	$Trap_5$	$Overlapping\,Trap_5$
Variables	100	99	100	100
Number of Evaluations	1522.566	15516.733	19932.633	72130
Number of Cliques	92.3	97.466	98.733	106.5
Size of Cliques	2.438	2.257	2.002	2.17
Edge number	163.6	140.133	99.266	137.133
Learning Index	0.318	0.466	0.663	0.392*
Similarity Index	0.449	0.772	0.396	0.354*
The Optimum value	80	99	100	110
Mean best value	80	99	100	109.933
Sample Complexity x	$x < 0.04$	$x < 0.08$	$x < 0.09$	$x < 0.09$

Table 15. AETC – EDA (CG - Sampler with a Metropolis step). Population size= 140. $T = 8000$. $\alpha = 0.9$.

Parameters	F_{c_2}	F_3	$Trap_5$	$Overlapping\,Trap_5$
Variables	100	99	100	100
Number of Evaluations	2551.166	24823.366	28188.766	22701.166
Number of Cliques	82.9	94.666	98.666	97.966
Size of Cliques	2.306	2.092	2.004	2.189
Edge number	134.933	110.233	99.433	129.933
Learning Index	0.467	0.494	0.668	0.458*
Similarity Index	0.483	0.703	0.4	0.402*
The Optimum value	80	99	100	110
Mean best value	80	99	100	110
Sample Complexity x	$x < 0.02$	$x < 0.04$	$x < 0.06$	$x < 0.05$

Table 16. AETC – EDA (CG - Sampler with a Metropolis step). Population size= 180. $T = 8000$. $\alpha = 0.9$.

Parameters	F_{c_2}	F_3	$Trap_5$	$Overlapping\,Trap_5$
Variables	100	99	100	100
Number of Evaluations	4854.133	28234	35599.4	28616.266
Number of Cliques	86	89.2	98.466	95.066
Size of Cliques	2.224	2.181	2.005	2.15
Edge number	124.566	116.366	99.566	117.566
Learning Index	0.582	0.528	0.669	0.535*
Similarity Index	0.497	0.775	0.402	0.41*
The Optimum value	80	99	100	110
Mean best value	80	99	100	110
Sample Complexity x	$x < 0.01$	$x < 0.03$	$x < 0.04$	$x < 0.03$

Table 17. AETC – EDA (CG - Sampler with a Metropolis step). Population size= 1000. $T = 8000$. $\alpha = 0.9$.

Parameters	F_{c_2}	F_3	$Trap_5$	$Overlapping\,Trap_5$
Variables	100	99	100	100
Number of Evaluations	202306.4	125055.033	17410.2	35843.866
Number of Cliques	98.966	96.566	97.133	97.2
Size of Cliques	2	2.015	2.02	2.022
Edge number	99.033	99.43	101.033	101.2
Learning Index	0.656	0.504	0.667	0.667*
Similarity Index	0.39	0.674	0.41	0.411*
The Optimum value	80	99	100	110
Mean best value	80	99	100	110
Sample Complexity x	$x < 0.0004$	$x < 0.0009$	$x < 0.001$	$x < 0.0009$

The two experiments give the following results: the number of evaluations differ between test functions, and between population sizes (see tables). The optimum is obtained in practical all the cases, except in only one run (see Table 14) for $Overlapping\,Trap_5$ function, where a local optimum is obtained. In all tables the * means, in the case of the $Overlapping\,Trap_5$ function, that the learning and similarity indexes are calculated with respect to the hypothetical model of $Trap_5$ function. The learning and similarity indexes, showed in the tables are calculated with the last model obtained where the optimum is found. However, intermediate models obtained along the evolution of the algorithm reach better learning and similarity indexes.

Observing the figures resuming the relationships between population size, and number of evaluations, it is seen that the algorithms performance for test function Fc_2 differs from the others test functions (see figures). The population size of 500, decreases the number of evaluations, canceling the effect of the Metropolis step, except for the Fc_2 function (see Figure 1). The performance of this test function is better for short population sizes, coinciding with the results founded in [13].

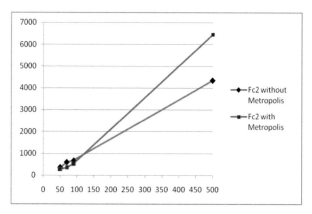

Fig. 1. Fc2 function

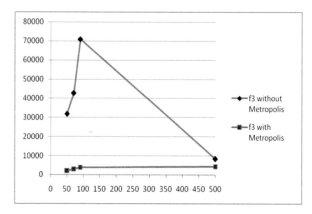

Fig. 2. F3 function

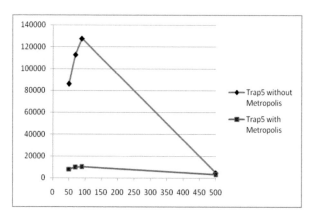

Fig. 3. Trap5 function

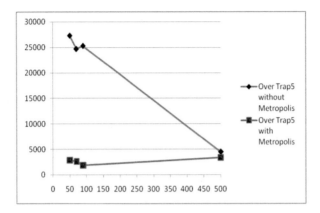

Fig. 4. Overlapping Trap5 function

The number of evaluations of the test functions F_3, $Trap_5$ and $OverlappingTrap_5$ (see Figures 2, 3, and 4) grows first and then go down at population size of 500. The number of evaluations of the algorithm using Metropolis step has a better performance for all the test functions employed than the algorithm without using it, except for the test function Fc_2.

7   Conclusions

As results of the experiments is to emphasize that: the algorithms obtain the optimum in practically all cases, the algorithm using the Metropolis step outperforms the other one in all cases, except for Fc_2 function, because it can be optimized with small population sizes either with Metropolis step or not. For 48 and 50 variables and populations of 500 individuals, models of cliques with 2 variables (a tree model) are sufficient to obtain the optimum, and the same result is obtained for 99 and 100 variables and populations of 1000 individuals. The algorithms use the learned cliques at each step of the inner cycles to obtain the best solution at hand. With growing population sizes the mean size of cliques employed by the algorithm with or without the Metropolis step arrives short to 2, but this arriving is faster with metropolis step algorithm. The sample complexity is always greater than zero and decreases proportionally to the population size. So, both, mean size of cliques and the sample complexity decrease proportionally to the population size. The same can be said for the mean number of edges of the graphical model. With respect to the number of evaluations of the algorithm, using Metropolis step, a better performance is observed for all the test functions employed, than without using it, except for the test function Fc_2. The similarity index for the algorithm with Metropolis step is less than the similarity index for the algorithm without Metropolis step. The learning index increases directly proportional to the population size. The reason could be that the algorithm has more information about the optimization problem when the size population is increased. It seems as if for this benchmark functions used linkage learning is more important (mean cliques sizes) for shorter populations, and for larger populations cliques of size 2 are enough. As future work, experiments using intermediate population sizes will be performed, and more complex test functions will be used.

Acknowledgements. We would like to acknowledge the support for this project (PIINF12-8) given by the Autonomous University of Aguascalientes, Aguascalientes, Mexico.

References

1. Brownlee, A.E.I., McCall, J.A.W., Shakya, S.K., Zhang, Q.: Structure Learning and Optimization in a Markov Network based on Estimation of Distribution Algorithm. In: Chen, Y.-P. (ed.) Exploitation of Linkage Learning. ALO, vol. 3, pp. 45–69. Springer, Heidelberg (2010)
2. Berge, C.: Graphs and Hypergraph. North-Holland, Amsterdam (1976)
3. Besag, J.E.: Spatial interaction and the statistical analysis of lattice systems (with discussion). J. Royal Statist. Soc. Series B 36, 192–326 (1974)
4. Chow, C.K., Liu, C.N.: Approximating discrete probability distributions with dependence trees. IEEE Transactions on Information Theory IT-14(3), 462–467 (1968)
5. De la Maza, M., Tidor, B.: An analysis of selection procedures with particular attention paid to proportional and Boltzmann selection. In: Proceedings of the 5th International Conference on Genetic Algorithms, pp. 124–131. Morgan Kaufmann Publishers Inc., San Francisco (1993)
6. Diaz, E., Ponce-de-Leon, E., Larrañaga, P., Bielza, C.: Probabilistic Graphical Markov model Learning: An Adaptive Strategy. In: Aguirre, A.H., Borja, R.M., Garciá, C.A.R. (eds.) MICAI 2009. LNCS, vol. 5845, pp. 225–236. Springer, Heidelberg (2009)
7. Etxeberria, R., Larrañaga, P.: Global optimization with Bayesian networks. In: Proceedings of II Symposium on Artificial Intelligence, CIMAF 1999. Special Session on Distributions and Evolutionary Optimization, La Habana, Cuba, pp. 332–339 (1999)
8. Geman, S., Geman, D.: Stochastic relaxation, Gibbs distributions and the bayesian distribution of images. IEEE Transactions on Pattern Analysis and Machine Intelligence 6, 721–741 (1984)
9. Holland, J.H.: Adaptation in natural and artificial system, pp. 11–12. Univ. of Michigan Press, Ann. Arbor (1975/1992)
10. Kindermann, R., Snell, J.L.: Markov random fields and their applications. American Mathematical Society, Contemporary Mathematics, Providence, RI (1980)
11. Kruskal, J.B.: On the Shortest Spanning Tree of a Graph and the Traveling Salesman Problem. Proceeding American Mathematical Society 7, 48–50 (1956)
12. Kullback, S., Leibler, R.A.: On information and sufficiency. Annals of Mathematical Statistics 22(1), 79–86 (1951)
13. Larrañaga, P., Lozano, J.A.: Estimation of Distribution Algorithms: A New Tool for Evolutionary Computation. Kluwer Academic Publishers (2002)
14. Lauritzen, S.L.: Graphical models. Oxford University Press, USA (1996)
15. Metropolis, N., Rosenbluth, A.E., Rosenbluth, M.N., Teller, A.N., Teller, E.: Equation of state calculations by fast computing machines. J. Chem. Phys. 21, 1087–1092 (1953)
16. Mühlenbein, H.: The equation for the response to selection and its use for prediction. Evolutionary Computation 5(3), 303–346 (1997)
17. Mühlenbein, H., Mahnig, T., Ochoa Rodriguez, A.: Schemata, Distributions and Graphical Models in Evolutionary Optimization. Journal of Heuristic 5(2), 215–247 (1999)

18. Mühlenbein, H., Mahnig, T.: FDA a scalable evolutionary algorithm for the optimization of additively decomposed functions. Evolutionary Computation 7(4), 353–376 (1999)
19. Mühlenbein, H., Paaß, G.: From recombination of genes to the estimation of distributions I. Binary parameters. In: Ebeling, W., Rechenberg, I., Voigt, H.-M., Schwefel, H.-P. (eds.) PPSN 1996. LNCS, vol. 1141, pp. 178–187. Springer, Heidelberg (1996)
20. Munetomo, M., Goldberg, D.E.: Linkage Identification by Non - monotonicity Detection for Overlapping Functions. Evolutionary Computation 7(4), 377–398 (1999)
21. Pelikan, M.: Bayesian Optimization Algorithm: From Single Level to Hierarchy. University Illinois at Urbana Champain, PHD Thesis. Also IlliGAL Report No. 2002023 (2002)
22. Pelikan, M., Mühlenbein, H.: The bivariate marginal distribution algorithm. In: Roy, R., Furuhashi, T., Chawdhry, P.K. (eds.) Advances in Soft Computing - Engineering Design and Manufacturing, pp. 521–535. Springer, London (1999)
23. Ponce-de-Leon-Senti, E., Diaz, E.: Adaptive Evolutionary Algorithm based on a Cliqued Gibbs Sampling over Graphical Markov Model Structure. In: Shakya, S., Santana, R. (eds.) Markov Networks in Evolutionary Computation. ALO, vol. 14, pp. 109–123. Springer, Heidelberg (2012)
24. Ponce-de-Leon-Senti, E.E., Diaz-Diaz, E.: Linkage Learning Using Graphical Markov Model Structure: An Experimental Study. In: Schütze, O., Coello Coello, C.A., Tantar, A.-A., Tantar, E., Bouvry, P., Del Moral, P., Legrand, P. (eds.) EVOLVE - A Bridge Between Probability, Set Oriented Numerics, and Evolutionary Computation II. AISC, vol. 175, pp. 237–249. Springer, Heidelberg (2012)
25. Santana, R., Mühlenbein, H.: Blocked Stochastic Sampling versus Estimation of Distribution Algorithms. In: Proceedings of the 2002 Congress on the Evolutionary Computation CEC 2002, pp. 1390–1395. IEEE Press (2002)
26. Shakya, S.: DEUM: A Framework for an Estimation of Distribution Algorithm based on Markov Random Fields. PhD Thesis (2006)
27. Shakya, S., Santana, R.: An EDA based on Local Markov Property and Gibbs Sampling. In: Proceedings of the 10th Annual Conference on Genetic and Evolutionary Computation, GECCO 2008, Atlanta, US, pp. 475–476. ACM Digital Library (2008)
28. Tsuji, M., Munetomo, M., Akama, K.: Linkage Identification by Fitness Difference Clustering. Evolutionary Computation 14(4), 383–409 (2006)
29. Van Kemenade, C.H.M.: Building Block Filtering and Mixing. In: Proceedings of the 1998 IEEE International Conference on Evolutionary Computation, pp. 505–510. IEEE Press (1998)
30. Winter, P.C., Hickey, G.I., Fletcher, H.L.: Instant Notes in Genetics, 2nd edn. Springer, New York (2002)

Support Tool for a Bayesian Network Based Critical Infrastructure Risk Model

Thomas Schaberreiter[1,2,3], Pascal Bouvry[2], Juha Röning[3],
and Djamel Khadraoui[1]

[1] Centre de Recherche Public Henri Tudor, Service Science & Innovation (SSI), 29,
avenue John F. Kennedy, L-1855 Luxembourg, Luxembourg
`thomas.schaberreiter@tudor.lu`
[2] University of Luxembourg, Computer Science and Communications Research Unit,
6, rue Richard Coudenhove-Kalergi, L-1359 Luxembourg
`pascal.bouvry@uni.lu`
[3] University of Oulu, Department of Electrical and Information Engineering,
P.O. Box 4500, FIN-90014 University of Oulu, Finland
`juha.roning@ee.oulu.fi`

Abstract. Critical infrastructures (CIs) provide important services to
society and economy, like electricity, or communication networks to enable telephone calls and internet access. CI services are expected to provide safety and security features like data *Confidentiality* and *Integrity*
as well as to ensure service *Availability* (CIA). The complexity and interdependency of CI services makes it hard for CI providers to guarantee
those features or even to be able to monitor the CIA risk by taking into
account that an incident in one CI service can cascade to another CI
service due to a dependency.

This work presents a tool implementing a previously published
Bayesian network based CI risk model which attempts to address the
challenges of interdependent CI risk monitoring. While Bayesian networks provide a great theoretical basis for CI risk monitoring, tool
support to cover the challenges in this field is missing. The tool was
implemented to provide visual guidance for domain experts to generate a CI risk model from real-world CIs and to simulate/emulate risk
scenarios based on this model.

Keywords: Critical infrastructures, Critical infrastructure modelling,
Bayesian networks, Dynamic Bayesian networks, Risk estimation, Risk
prediction.

1 Introduction

Critical infrastructure (CI) security has become an important research topic in
the last years. CIs are service providers, the services they provide are so vital to
the social and economic well-being of a society that a disruption or destruction of
the infrastructure would have severe consequences. CI sectors include, amongst
others, the telecommunication, electricity and transport infrastructures.

O. Schütze et al. (eds.), *EVOLVE - A Bridge between Probability, Set Oriented Numerics,*   53
and Evolutionary Computation III, Studies in Computational Intelligence 500,
DOI: 10.1007/978-3-319-01460-9_3, © Springer International Publishing Switzerland 2014

Security in CIs can be seen from many viewpoints. In this work, the term CI security (or CI risk) is used to define the risk of a breach of confidentiality, the risk of a breach of integrity and the risk of degrading availability (CIA) of a CI or CI service. During our research some key factors that influence CI security were identified: First of all, the *complexity* of CIs makes security complex. CIs are large organizations and their security is influenced by technical, social and organizational factors both at a national and international level. This complexity makes it hard to identify the most critical parts of a CI and their interactions which influence CI security. Secondly, the *dependencies* and *interdependencies* between CIs and CI services influence the security of CIs. CIs can be dependent on the service of other CIs and a security incident in one CI or CI service can cascade and cause an incident in another CI or CI service. Interdependencies exist when an incident cascades back to the initial service through dependent services. Thirdly, the *diversity* of CI sectors makes it hard to have a holistic view on CI security, e.g. by including dependencies to other CI sectors in CI risk estimation.

CI security modelling ([2], [1]) was introduced to address those problems and provide a CI model for on-line risk monitoring. The main entities of the CI security model are *CI services* and *dependencies* between CI services. The model can estimate the CI service risk after a security incident (observed by *system measurements*) and distribute this estimate to dependent CI services which in turn can use this information to update their risk estimate. The estimated CI service risk level is represented by a risk value between one (no risk) and five (maximum risk). The strong points of the CI security model are that by using CI service risk as the model output, uniform and comparable data is created that is valid for all CI sectors, and that dependencies can be included in CI service risk estimation. In [13] this model was extended with a Bayesian network based component to achieve a more powerful approach for risk estimation which also allows some advanced features like *risk prediction* and handling of *interdependencies* using Dynamic Bayesian networks (DBNs).

The contribution of this work is the presentation of a tool based on the ideas published in [13]. The main tasks of the tool are on the one hand to provide a simple and easy to understand graphical user interface to support the decomposition of a CI into the CI security modelling structure (CI services, dependencies and base measurements), to facilitate expert input and to provide a simple visual way for base measurement normalization and initial estimation of CI service risk (those steps will be detailed further in Section 3). On the other hand the tool provides an implementation of algorithms for automatic CI service risk probability learning. Although several general-purpose Bayesian network tools are available, non of those tools fulfilled the requirements of seamless integration with the presented tool and were not able to handle some specific requirements of CI security modelling like CI service risk prediction and handling of interdependencies using DBNs. The presented tool fully supports the concepts of building a Bayesian network based CI security model presented in [13] as well as risk simulation/emulation in such a model.

The remainder of the paper is organized as follows: In Section 2, work related to CI security modelling is discussed. In Section 3, relevant ideas of CI security modelling are presented to set the context for the tool described in Section 4. Finally, Section 5 concludes the paper and gives an outlook on future work.

2   Related Work

The concept of CI security modelling relates to several research areas: CI modelling and simulation, CI (inter)dependency identification and risk estimation in CIs. Identifying the various kinds of dependencies among CIs has been subject to previous research. In [10] Rinaldi et al. provide an excellent overview on the dimensions where interdependencies can occur. Several publications propose CI models based on various different modelling techniques. For example, conceptual modelling is used in [16] by Sokolowski et al. to represent an abstract and simplified view of CIs. In [8] Panzieri et al. utilise the complex adaptive systems (CAS) approach for CI modelling. The model is derived by modelling the mutually dependent sub-systems of the infrastructure. Risk models for CIs were proposed by some authors. For example, in [7] Haslum et al. use continuous-time hidden Markov models for real-time risk calculation and estimation. In [3] Baiardi et al. propose a risk management strategy based on a hyper-graph model to detect complex attacks as well as to support risk mitigation. In [6] Haimes et al. propose an eight step risk ranking and filtering framework based on risk scenarios, using hierarchical holographic modelling. Bayes theorem is used to estimate the likelihood of risk scenarios. In general, previously published CI models and CI risk models vary greatly in their purpose and the extent to which they were implemented. The models are usually too high-level and therefore lack practical relevance or they are focused on a specific CI and therefore lack generality.

The idea of the CI security modelling differs greatly from the models previously published. It tries to establish abstract models of CIs that can be compared with each other while maintaining generality by enabling it to be applied to all kinds of CI sectors. CI security modelling is ongoing research, the first publications date from 2010 ([2], [1], [11]). The intend of those publications was to introduce the service based notion of CIs and to illustrate the dependence of services amongst another. The need of a common modelling entity to address the diversity of different CI sectors was introduced by using a risk-based CI model. Risk is estimated from system measurements using a weighted-sum method. One of the shortcomings of this original model is that it relies heavily on expert assessment. Despite evaluating the structure and dependencies of a CI, experts also assess the importance (weight) of each system measurement to a service in order to be able to calculate risk using a weighted sum. In [13] BNs are introduced to the CI security model to provide a more sophisticated yet more convenient way of representing risk which allows to address shortcomings of the original proposal. Using this approach, system measurements do not need to be weighted for their importance which reduces the dependence on expert knowledge. Using BNs and their temporal extension, DBNs, as a modelling base also allows to include some features that were not covered by the original proposal, like risk prediction or a way to model interdependencies.

To address the complexity of CIs and to reduce the dependence on expert assessment in determining the structure of the CI security model, a CI analysis method based on dependency analysis was presented in [15]. The idea is to utilise different information sources on different organizational levels and to combine social as well as technical sources to get a holistic view on the investigated CI and to identify critical services, dependencies between critical services and system measurements to determine CI service states.

In [14] a set of indicators is presented that allow to evaluate the correctness of calculated service risk in form of an assurance indicator and in [4] the risk indicators received from dependent services are evaluated based on the estimated trust in those services.

A multi-agent based tool supporting risk monitoring using the CI security modelling approach was presented in [12]. This tool is based on a multi-agent platform which allows it to operate in distributed environments. The objective of the tool is to support the deployment phase of a CI security model where parts of the monitoring environment need to run independently in different locations (e.g. in different CIs or different sites of a CI). The tool presented in this work on the other hand does not cover the deployment phase of the CI security model, but the design of the model. Although the two tools are not compatible yet, in future CI security models created by this tool could be used to configure the tool presented in [12].

Although the previously published work related to CI security modelling represents a solid theoretical basis, the practical relevance could not be evaluated due to the lack of tool support. The work presented in this article addresses this issue by presenting a tool that fully supports the concepts of Bayesian network based CI security modelling presented in [13].

3   BN based Critical Infrastructure Security Model

BNs provide a way to model the probability of an event, given the state of events it depends on. BNs are directed acyclic graphs where the nodes represent event variables and the directed arcs represent the relationships or dependencies between the variables. Nodes in a BN are only dependent on their parent nodes, not on any other ancestors (conditional independence). In other words, conditional independence is given if a node is independent of its ancestors given the state of its parents. Each node in the BN has a conditional probability table (CPT) assigned to it containing the probabilities of the node being in a certain state, given the state of the parent nodes. Each possible combination of each state of the parent nodes has to be evaluated. It can be easily seen that the CPTs for nodes with a significant amount of parent nodes with multiple states are quite complex. The size of a CPT grows exponentially with the number of parent nodes.

In this work the main objective of the BN is to be used as a classifier. In this case, one is interested in the most probable state of a variable given a combination of states of the parent variables. For example, it is observed that variable A is in state "true" and variable B is in state "false". What is the most probable state of variable C?

Several reasons lead to using BNs for this model. First of all, the graphical structure of BNs is very intuitive and easy to interpret for people not familiar with the modelling domain, no model specific aspects need to be considered. Secondly, BNs have the capability of learning probabilities from data as well as being able to be assigned by experts. It is assumed that it is essential in the CI field, where sophisticated expert knowledge is available, experts will evaluate probabilities that can not be learned from data or re-evaluate probabilities that were learned from insufficient data. Furthermore, BNs provide a very natural way to model dependencies. One of the main goals of BNs is to capture the relationships and dependencies between events, as does the CI security model.

3.1   Structure for Bayesian Network

When building a model of complex systems, understanding the structure of the system and the interaction between components and other external systems is crucial. It is equally important to understand the purpose of the model to be able to map the real system to the abstract entities of the model.

In the CI security model the central modelling entities are *CI services* and the interactions (or *dependencies*) between services. A CI service is provided by a CI either to customers or to other CI services as a dependency. The main objective is to model CI service risk by observing *system measurements* that define the service state and by observing dependent CI service risk. The mapping between real-world observations and abstract risk estimates is done by the Bayesian classifier.

The main concern when building a CI model is complexity. Is it feasible to identify the critical services and dependencies that adequately represent the structure in a complex system like a CI and to represent them in a model? In an attempt to address this question a CI analysis method based on dependency analysis was proposed in previous work ([15]). This method adapts the PROTOS-MATINE dependency analysis method ([5], [9]) to be used to find the structure for the CI security model and to identify the modelling entities used in the CI security model (critical services, dependencies and system measurements). The method is based on the assumption that in order to get a holistic view of a complex system, all available information sources (e.g. documentation, manuals, interviews, contracts, ...) on all levels in an organization (management, process, technical) need to be combined. The outcome of this method is a graph that contains the critical services as nodes and their dependencies as edges. System measurements that define a CI service state can be seen as dependencies of this service. An example for the structure of a CI security model can be seen in Figure 1. The nodes S_A to S_D represent critical services, the nodes $M1$ to $M7$ represent system measurements. The edges represent a dependency between services (e.g. the availability of S_A depends on S_B) or between a service and a system measurement (e.g. the availability of S_A depends on the state of M_1).

58   T. Schaberreiter et al.

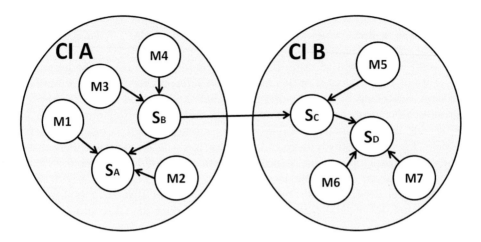

Fig. 1. Example of a CI security model structure

The grey cycles CI A and CI B do not have any purpose in the model, they should illustrate that dependent services can also belong to different critical infrastructures (e.g. S_B of CI A is needed to ensure availability of S_C in CI B).

3.2 Conditional Probability Tables

After evaluating the structure of the BN, the next step is to evaluate the CPTs for each node. As mentioned before, the BN is used as a classifier and the main interest is in determining the most probable state of a node for each combination of states of the parent nodes.

The following list contains a summary of definitions and pre-conditions to be able to learn the conditional probability tables for the nodes in the context of the CI security model:

1. The state of a CI service node represents risk (for example, C, I or A). The risk value is limited to 5 discrete states, with 1 representing lowest risk and 5 representing highest risk. Since service risk is an abstract concept and can not be directly measured, an expert has to evaluate the risk a service experienced during the time period used to learn the probabilities (for example, an expert estimates that from time x to time y the service faced an incident that can be classified confidentiality risk level 3). With this information, the probabilities with respect to the states of the parent nodes can be learned from recorded data samples.
2. System measurement usually represent continuous measurements. In order to be more easily processed, they are pre-processed to 5 discrete values with 1 representing a measurement during normal system operation and 5 representing a maximum allowed deviation from normal operation. Boolean-type measurements will only have the states 1 (normal operation) and 5 (abnormal operation).

3. System measurements need to be observable and it is assumed that their values can be recorded over time and can be combined with a time stamp.
4. Each CPT represents the probabilities for one risk indicator (either C, I or A). If more than one risk indicator is interesting for a CI service, probabilities for each indicator have to be estimated separately.

Learning of conditional probabilities for CI service CPTs is a matter of finding each occurrence of a certain dependency state combination (where a dependency can be a base measurement or another CI service) in the data set and determining the service risk state at the time of the occurrence. The most frequent service risk state is the most probable service risk if this dependency state combination occurs.

Probabilities that can not be learned from data samples because of incomplete data (state combinations that never happened or only rarely happened) can be supplemented by experts. This can be a burden since CPTs might be complex (CPTs grow exponentially with the number of parents of a node).

3.3   Risk Prediction

The CPTs described in the previous section only capture the current risk of a service (if there is a certain combination of parent node states, what is the most probable risk of the service). In practice companies are usually also interested in the evolution of risk over time after an event occurred. One way of representing this is to estimate the *short-term* (e.g. hours after an event) *mid-term* (e.g. weeks after an event) and *long-term* (e.g. months after an event) effects of an event or incident. In the BN model this can be represented using DBNs. DBNs are an extension of BNs to allow to model changing temporal relationships between variables. Both dependencies between variables and conditional probabilities can change over time. DBNs can model this by representing each time frame t by a separate BN and linking the time slices in the direction of the time flow.

The basic idea of CI service risk prediction is to separate the data that is used to learn the BN into time frames after an event happened. A CPT can be learned (or estimated by an expert) for each time frame. This will give an estimate of the most probable state of a service in each time frame, e.g. if an incident happens, what is the risk the service faces in the next hours, the next weeks and the next months.

Using DBN for risk prediction makes learning the CPTs considerably more complex. More CPTs need to be considered since each node has a separate CPT in each time frame and the amount of data needed to learn grows in terms of the time that has to be considered as well as the amount of recorded incidents to be able to capture all possible states in each time frame.

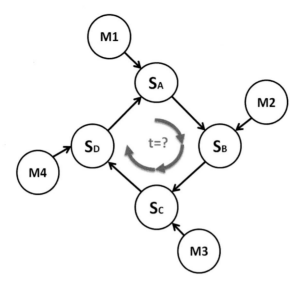

Fig. 2. Example of a cycle in a BN

3.4 Interdependencies – Directed Cycles in Bayesian Networks

One of the shortcomings of BN models is that it is not easily possible to model directed cycles. In the context of the CI security model this is a problem, since one of the main goals of the work is to model interdependencies between CI services. Interdependencies exist when an incident in one CI service effects the same service again through dependent services. An illustration of this behaviour can be seen in Figure 2, where the dependency cycle is $S_A \rightarrow S_B \rightarrow S_C \rightarrow S_D \rightarrow S_A$. The nodes $M1, M2, M3, M4$ represent system measurements that can change the state of the service nodes.

One way to address the shortcoming of BNs for this scenario is to use DBNs. The idea is to estimate the time (t) it takes for an incident to loop back to a service through a dependent service. This time is taken as the time frame to build the DBN. The CPT in the first time frame represents the probabilities for CI service risk given an event without the loop-back effect, the CPT in the second time frame would represent the probabilities of service risk under the assumption that the effects of the original event loop back during this time frame. The CPTs of the third, fourth and n-th time frame represent the probabilities of service risk after the second, third and (n-1)th loop, respectively.

4 Support Tool

In this section the tool that was implemented to support the Bayesian network based critical infrastructure risk model is presented. The purpose of the tool is to graphically support all the steps of the CI risk model in an intuitive way so that expert input is facilitated. Furthermore, the tool allows to simulate or emulate CI risk behaviour after the model is completed.

The design principles for the graphical user interface (GUI) were to provide a *clean, well structured* interface to the user that is *easy to understand and use*, and contains all the elements of the Bayesian network based CI security model. To achieve those goals, the tasks were split into 5 separate sub-tasks which are realized in 5 separate tabs in the GUI of the tool:

1. **CI structure:** Decomposition of CI and representation in services, dependencies and base measurements.
2. **Data pre-processing:** For each service, allow to enter base measurement normalization bounds and service risk estimation.
3. **Interdependencies:** Allow special handling of interdependencies if cycles in the graph structure exist.
4. **Conditional probability tables:** For each service, allow to learn conditional probabilities and allow expert estimation where incomplete or insufficient data is available to learn the probabilities.
5. **CI simulation/emulation:** Allow to simulate or emulate CI risk behaviour and service risk propagation through dependencies.

Those 5 tasks will be detailed in the following sections.

4.1   CI Structure

The structure of the CI according to the CI security model (services, dependencies between services and base measurements) needs to be determined using a dependency analysis method like the one presented in [15]. To graphically support this analysis, the CI structure tab of the tool is designed to create a graph that is composed of those elements, as can be seen in the screen-shot in Figure 3. The incentive of this representation is to provide an easy to understand overview of the structure and identify possible flaws and errors in the representation of the CI security model at an early stage.

Each service and base measurement in the model is given a unique name upon creation, dependencies are created via drag-and-drop between two elements. To be able to learn probabilities at a later stage, each base measurement needs to be assigned a file containing recorded data samples where each sample has the format "Unix time stamp"[1];"value" (e.g. 1338527100;0.002). Each service can be assigned a service risk file that contains experienced service risk at certain times in the same format as base measurement files, with the difference that service risk values can only be 1,2,3,4 or 5. Since service risk is an abstract concept that can not be directly measured, this information might not be available at

---

[1] A time stamp in Unix time format represents the number of seconds since the the 1st of January 1970 UTC (Coordinated Universal Time). For example, 1338527100 is "Fri, 01 Jun 2012 05:05:00" UTC.

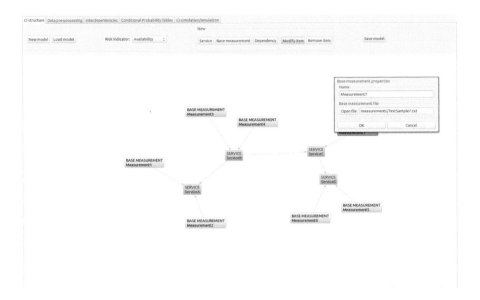

Fig. 3. Screen-shot of CI structure tab

the time of creation. Therefore the service risk file association is not mandatory and can be estimated at a later stage.

To account for the fact that each risk indicator (Confidentiality, Integrity and Availability - CIA) can have a different security model structure, each risk indicator model is created separately. Using the *Risk Indicator* drop-down menu in the top panel of the tab, the appropriate risk indicator model can be selected. Although not necessary, each indicator model can have the same services and base measurements (which means that the same CI service might or might not want to monitor the risk for each indicator). Therefore, services and base measurements with the same name in the different indicator models will be treated as one service at a later stage.

To be able to resume working on a CI security model, the tool allows to save the model to a XML-based file structure. Aside from the security model structure, any information provided in one of the subsequent steps will be stored using this file structure.

4.2   Data Pre-processing

The data pre-processing tab is designed for two purposes: To estimate the CI service risk where no service risk file was provided and to set the base measurement normalization bounds for each base measurement a service depends on. This information is needed to be able to learn conditional probabilities from data at a later stage. To show the data for a service, the appropriate service can be selected using the *Risk indicator* and *CI services* drop-down menus.

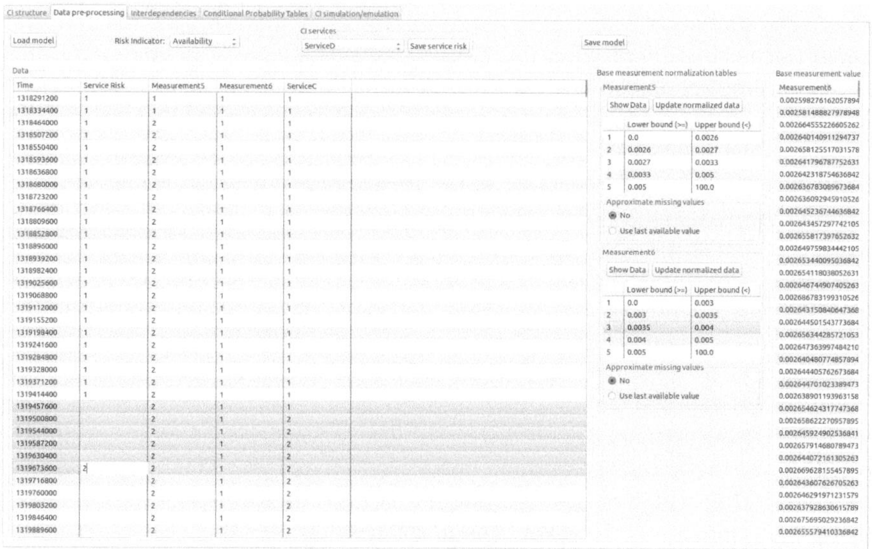

Fig. 4. Screen-shot of data pre-processing tab

As a first step, all available time stamps found in the base measurement files for the selected indicator will be displayed in the *Time* column of the *Data* section. The *Service risk* column of the *Data* section represents the service risk of the selected service at the displayed time points. If a service risk file was provided, this column is filled with the contents of the file. If not, this column remains empty. It is editable so that experts can provide service risk estimates for the corresponding time points. It is necessary to load the time stamps of all base measurements of an indicator model and not only the time stamps from the base measurement the current service depends on, because the service risk estimation is used by dependent services for learning conditional probabilities. But if the base measurement files of those dependent services contain more data samples, there will be no service risk estimate and those data samples can not be used. As an example, considering the model displayed in Figure 3, *ServiceD* depends on *ServiceC*. If *Measurement7* contains less data samples than *Measurement5* or *Measurement6*, service risk of *ServiceC* would not be estimated if only time stamps for *Measurement7* would be considered. Therefore, the additional data samples of *Measurement5* and *Measurement6* could not be used for learning conditional probabilities.

The rest of the columns in the *Data* section are base measurements and services the current service depends on. If the dependency is a service, the corresponding service risk file is displayed. If not, the service risk estimation for this service has to be done first to be able to complete the data. If the dependency is a base measurement, a table to enter normalization bounds will be displayed in the *Base measurement normalization tables* section so that continuous measurements can be normalized to 5 steps, as specified by the security model. The tool allows to set the lower and upper bound for each discrete step as well as to choose approximation if no value is available at a time point. The approximation was introduced to account for small discrepancies in the time a measurement was taken. For example, assuming that the interval a measurement is taken is 30 seconds for both *Measurement5* and *Measurement6*, but *Measurement6* measures 15 seconds later than *Measurement5*, there would never be a measurement taken at the same time and learning conditional probabilities would not be possible. Setting the approximation method to *Use last available value* for a base measurement fills gaps at time points where no value is available with the last observed state. This method is obviously not suited if there are large gaps between subsequent measures, but sufficiently approximates small gaps.

If the *Show data* button for a base measurement is pushed, the original values of this base measurement are shown in the *Base measurement values* section to visually assist in finding the optimal normalization bounds. The *Update normalized data* button will show the normalized data in the corresponding column of the *Data* section as well as saving the normalization bounds.

4.3   Interdependencies

To account for the possibility of interdependencies (directed cycles in the CI security model), the interdependencies tab as shown in Figure 5 was introduced. It is used to obtain information needed for special handling of services that are part of a cycle so that learning of conditional probabilities using dynamic Bayesian networks can be achieved. For each indicator model, which can be selected via the *Risk indicator* drop-down menu, the tool identifies any cycles in the graph model and populates the *Interdependencies* drop down menu. For now, only simple cycles (each service can only be part of one cycle) are supported by the CI security model.

By selecting an interdependency, the effected services as well as the dependencies that form the cycle will be displayed on the right side and on the left side a section to enter the interdependency information is displayed. The required information is:

- **Round-trip time (in Seconds):** The estimated time an incident needs to loop back to the originating service.
- **#Round trips:** For each service it can be decided for how many loops service risk estimation is desired. In many cases, the main effects of the loop-back will be in the lower order loops, but making the number of considered loop-backs configurable makes the model more flexible.

4.4   Conditional Probabilities

The conditional probabilities tab, as shown in Figure 6, is designed to learn the conditional probability tables for each service, and allow for expert estimation

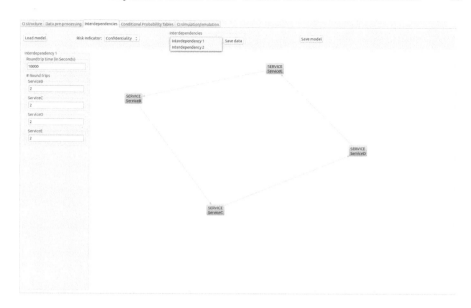

Fig. 5. Screen-shot of interdependencies tab

where learning is not possible due to insufficient data. Furthermore, the conditional probabilities tab is designed to allow to enable or disable risk prediction for each service as well as special handling of interdependencies, if cycles exist in the CI security model.

Using the *Risk indicator* and *CI services* drop-down menu allows to select the desired CI service, which will display the dependencies of the service as well as all possible input state combinations in the *Dependency states* section. On the right next to the *Dependency states* section, the CPTs for *Service risk probabilities*, *Short-term risk probabilities*, *Mid-term risk probabilities* and *Long-term risk probabilities* are shown. The risk prediction CPTs are only enabled if risk prediction is enabled.

By enabling the *Risk prediction* check button for a service in the *CI services section*, text fields in the *Time period after an event* section are enabled that allow to specify, in Seconds, what time period after an event happened is considered as short-term risk, mid-term risk and long-term risk. This information is used by the conditional probability learning algorithm.

By pressing the *Learn Conditional Probabilities* button for a service in the *CI services* section, the conditional probabilities will be learned using the data provided by the data pre-processing step. The following simple and straightforward algorithms are used to learn the conditional probabilities from data:

- **Learning of service risk probabilities:** For each occurrence of a dependency state combination, the associated service risk (1,2,3,4 or 5) is counted and at the end divided through the total number of occurrences of this dependency state combination, multiplied by 100%. This will provide the risk

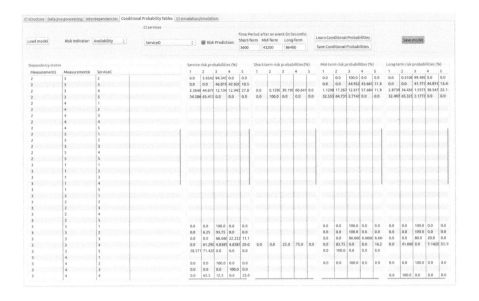

Fig. 6. Screen-shot of conditional probabilities tab

probability for this dependency state combination which is displayed in the
Service risk probabilities table.

- **Learning of risk prediction probabilities:** The risk prediction learning
  algorithm uses time to measure the changes in CI service risk after an event
  happened. For each occurrence of a dependency state combination, the algorithm will take the data samples recoded afterwards, within the time span
  provided by the *Time period after an event* text fields for short-, mid- and
  long-term prediction[2]. The service risk from those data samples in the respective time frames is counted and at the end divided through the maximum
  number of collected data samples in the time frame multiplied by 100%. This
  gives a probabilistic measure of service risk after an event with a certain dependency state combination happened, shortly after an event as well as in
  the mid- and long-term after the event.

After the learning algorithms finished, a window providing statistics from the
learning algorithm, as shown in Figure 7 will be displayed. For each dependency
state combination it is shown how many data samples with a certain risk state
were observed, as well as the total number of observed data samples for this
dependency state combination. Those statistics are shown for service risk probabilities as well as short-, mid- and long-term risk probabilities, if risk prediction
was enabled. The learning statistics can be saved to disk, if necessary.

---

[2] Taking the values from Figure 6, the timespan would be 0-3600 Seconds after the
event for short-term prediction, 3601-43200 Seconds after the event and 43201-86400
Seconds after the event.

Fig. 7. Screen-shot of conditional probability learning statistics

If risk probabilities for a dependency state can not be learned because this combination did not occur in the data used to learn the probabilities, those lines in the risk probability tables remain empty as can be observed in Figure 6. They can be supplemented with probabilities estimated by an expert. Please note that Bayesian networks are used as a classifier in this application and therefore only the risk state with the highest probability is of interest.

Services that are part of an interdependency need special handling to be able to learn the conditional probabilities for each round. As can be seen in Figure 8, each interdependency round that is considered by the model is treated separately and shown as a separate service in the *CI services* drop-down menu. For example, it can be seen in Figure 5 that for *ServiceD* in the Confidentiality risk indicator 2 interdependency rounds are considered. The *CI services* drop-down menu of the Confidentiality indicator in Figure 8 contains two instances of ServiceD: *(I1):ServiceD* and *(I2):ServiceD*, representing the probabilities of *ServiceD* without loop-back effect (I1) and with first-order loop-back effect (I2). Following conditional probability algorithm learning is used for learning conditional probability tables of services that are part of an interdependency:

- **Learning of interdependent service risk probabilities:** When the *Learn conditional probabilities* button of a service is pressed which is part of an interdependency, the time interval of the currently considered round (e.g. (I1) or (I2)) is calculated from the provided interdependency round-trip time. For example, using the round-trip time in Figure 5, (I1) would have a time interval of 0-10000 Seconds and (I2) would have a time interval of

10001-20000 Seconds. After that, the data set that is used to learn the conditional probabilities for an interdependency round is composed as follows: The algorithm iterates through the dataset until a data sample indicates that the service operates under normal condition (dependency state of all dependencies is 1). All data samples after this event that are in the desired time frame (e.g. (I1) or (I2)) are copied to a new data set. If the system goes to normal condition before the time frame ends, the algorithm is reset and starts a new time frame. The new dataset is used to learn the conditional probability table using the algorithms presented before. Risk prediction can be enabled individually for each interdependency round of a service, but obviously learning results can only be provided if the round-trip time is longer than the prediction interval of short-, mid- or long-term prediction.

Fig. 8. Special handling interdependent services

To illustrate the applicability of the presented conditional probability learning process, a practical example of the learning process without taking into account interdependencies and risk prediction, is shown. This small example is composed of a CI service where the CI service risk is estimated from two base measurements. The CI service risk represents the risk experienced by supporting services (like ssh, ping, mail server,...) of a computing cluster. The base measurements represent the data records of a service monitoring tool (Nagios) and a service configuration and deployment tool (Puppet). The data sets are data records from a real-world distributed computing cluster called Grid'5000 (http://grid5000.fr/).

The Nagios status base measurement data set is presented in Figure 9. The status information was parsed from log files and the textual state explanation (OK, WARNING, UNKNOWN, CRITICAL) was mapped to numeric states (1,2,3 and 5) to be able to visualize the data set. The dataset contains 4356 data points collected over the time span of one month. It can be observed that the normalization states 1 and 2 occur substantially more frequently than the states 3 and 5 which suggests that automatic learning for the high CI service risk state probabilities will be less accurate and will have to be supplemented by expert estimation.

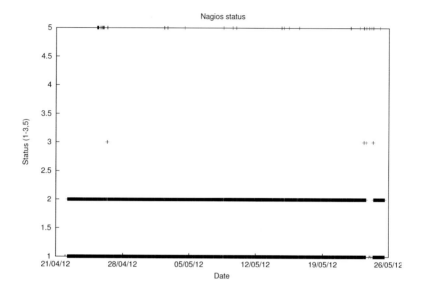

Fig. 9. Nagios status dataset

The Puppet status base measurement is visualized in Figure 10. The status information was parsed from log files and the textual state explanation (unchanged, changed, unresponsive, pending and failed) was mapped to numeric states (1-5) for visualization. The dataset contains 28987 data points collected over the time span of three months. The majority of data points are located in normalization states 1 and 2, with very few data points indicating failed deployments in normalization state 5. Again, those data points will not be sufficient to learn the risk probabilities for high CI service risk states.

The risk probabilities for the *Service Nodes* CI service are visualized in Figure 11. The probabilities were automatically learned using the previously preprocessed data sets. The learning statistics of this learning process are presented in Table 1. Only some of the probabilities, mainly in dependency state combinations containing the normalization values of 1 and 2, but also dependency state combinations where the Nagios status is 5, could be correctly learned. The other state combinations, which rarely occur in the data sets, could not be learned at all or were wrongly classified. The normalization state 4 is not a valid state for the Nagios base measurement and therefore the risk probabilities for dependency state combinations containing a Nagios state 4 are set to 0 for all possible CI service risk states. The learned risk probabilities were reviewed by the CI expert and all risk probabilities that could not be learned or were wrongly classified were manually re-estimated.

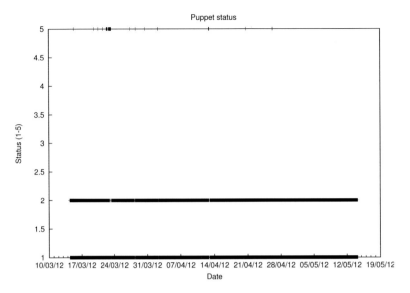

Fig. 10. Puppet status dataset

This simple illustration shows that learning of risk probabilities from data is possible and that the graphical representation by the tool allows estimation of risk probabilities where automatic learning is not possible.

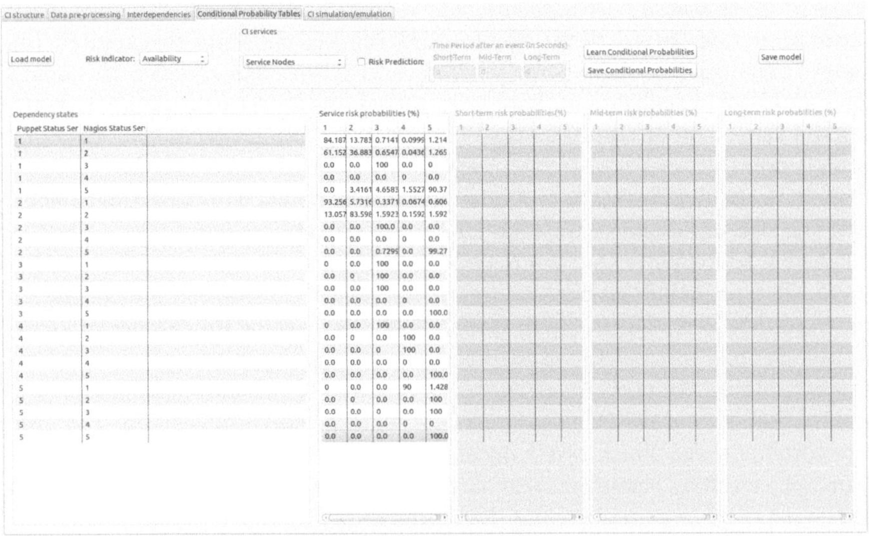

Fig. 11. Risk probabilities for Service Nodes CI service

Table 1. Risk probability learning statistics for Service Nodes CI service

State: 1,1	Number of occurrence: (1): 5894, (2): 965, (3): 50, (4): 7, (5): 85 Total number of occurrence: 7001
State: 1,2	Number of occurrence: (1): 1401, (2): 845, (3): 15, (4): 1, (5): 29 Total number of occurrence: 2291
State: 1,3	Number of occurrence: (1): 0, (2): 0, (3): 16, (4): 0, (5): 22 Total number of occurrence: 38
State: 1,4	Number of occurrence: (1): 0, (2): 0, (3): 0, (4): 1, (5): 0 Total number of occurrence: 1
State: 1,5	Number of occurrence: (1): 0, (2): 11, (3): 15, (4): 5, (5): 291 Total number of occurrence: 322
State: 2,1	Number of occurrence: (1): 4149, (2): 255, (3): 15, (4): 3, (5): 27 Total number of occurrence: 4449
State: 2,2	Number of occurrence: (1): 82, (2): 525, (3): 10, (4): 1, (5): 10 Total number of occurrence: 628
State: 2,3	Number of occurrence: (1): 0, (2): 0, (3): 43, (4): 0, (5): 0 Total number of occurrence: 43
State: 2,4	Number of occurrence: (1): 0, (2): 0, (3): 0, (4): 5, (5): 0 Total number of occurrence: 5
State: 2,5	Number of occurrence: (1): 0, (2): 0, (3): 1, (4): 0, (5): 136 Total number of occurrence: 137
State: 3,1	Number of occurrence: (1): 657, (2): 0, (3): 0, (4): 0, (5): 0 Total number of occurrence: 657
State: 3,2	Number of occurrence: (1): 0, (2): 44, (3): 0, (4): 0, (5): 0 Total number of occurrence: 44
State: 3,3	Number of occurrence: (1): 0, (2): 0, (3): 0, (4): 0, (5): 0 Total number of occurrence: 0
State: 3,4	Number of occurrence: (1): 0, (2): 0, (3): 0, (4): 0, (5): 0 Total number of occurrence: 0
State: 3,5	Number of occurrence: (1): 0, (2): 0, (3): 0, (4): 0, (5): 29 Total number of occurrence: 29
State: 4,1	Number of occurrence: (1): 962, (2): 0, (3): 0, (4): 0, (5): 0 Total number of occurrence: 962
State: 4,2	Number of occurrence: (1): 0, (2): 684, (3): 0, (4): 0, (5): 0 Total number of occurrence: 684
State: 4,3	Number of occurrence: (1): 0, (2): 0, (3): 22, (4): 0, (5): 0 Total number of occurrence: 22
State: 4,4	Number of occurrence: (1): 0, (2): 0, (3): 0, (4): 8, (5): 0 Total number of occurrence: 8
State: 4,5	Number of occurrence: (1): 0, (2): 0, (3): 0, (4): 0, (5): 134 Total number of occurrence: 134
State: 5,1	Number of occurrence: (1): 345, (2): 0, (3): 0, (4): 0, (5): 5 Total number of occurrence: 350
State: 5,2	Number of occurrence: (1): 0, (2): 152, (3): 0, (4): 0, (5): 4 Total number of occurrence: 156
State: 5,3	Number of occurrence: (1): 0, (2): 0, (3): 4, (4): 0, (5): 0 Total number of occurrence: 4
State: 5,4	Number of occurrence: (1): 0, (2): 0, (3): 0, (4): 2, (5): 0 Total number of occurrence: 2
State: 5,5	Number of occurrence: (1): 0, (2): 0, (3): 0, (4): 0, (5): 32 Total number of occurrence: 32

4.5   Risk Simulation/Emulation

The *Risk Simulation/Emulation* tab, as Seen in Figure 12, is used to provide a simple way to test the Bayesian network based security model and the conditional probabilities generated in the previous steps and to visualize the service risk propagation as a reaction to changing base measurement values. The differentiation between simulation and emulation means that using emulation, all base measurements are shown as editable text fields in the GUI to be changed by the user during emulation whereas simulation means that those values are read from a simulation file provided by the user[3] to ensure reproducible simulation results.

In the GUI, the CI security graph model for all indicators is displayed, if a service or base measurement is present in more than one indicator model, this is visualized by labelling the dependencies accordingly. When a service is clicked, risk plots for *current risk, short-term risk, mid-term risk* and *long-term risk* are displayed in the *Risk plots* section. The x-axis of the plots represents the time an event happened and the y-axis represents the service risk at that time. Only the risk indicators that are relevant for a service are displayed. For example, *ServiceD* in Figure 12 only considers risk prediction for the Availability indicator, but the service risk is estimated for Confidentiality and Integrity as well. Interdependency risk for services is visually represented by providing an *Interdependency* drop-down menu for services that are part of interdependencies. The risk for each interdependency round can be individually selected and displayed.

Each simulation or emulation starts with all services in normal operation (service risk for all services is 1). To account for this in emulation mode, all base measurements displayed in the *Base measurement values* section are set to an initial state *"Initial"*. Changing a base measurement value in simulation or emulation mode will result in re-calculation of service risk of all dependent services using the base measurement normalization tables shown in Figure 4 and the CPTs of the services. The risk is recursively updated for all dependent services, the risk plots are updated and all updated services are highlighted.

At the end of a simulation or emulation the results can be saved to disk. The results contain, for each base measurement, a list of changed values with time stamp and for each service a textual version of the risk plots.

4.6   Technical Details

The tool was implemented using Java programming language (compatible to Java version 7). Oracle Java 7 SDK (Software Development Kit) was used during implementation. Additional open-source libraries used for implementation are *Zest* (http://www.eclipse.org/gef/zest/) for drawing graphs in the GUI, *jgrapht* (http://jgrapht.org/) for determining simple cycles in graphs, *jfreechart* (http://www.jfree.org/jfreechart/) to display the risk plots and

---

[3] A line in a simulation file has the form $@t : X = x : Y = y$, which means that t Seconds after simulation start, base measurement X takes the value x and base measurement Y takes the value y.

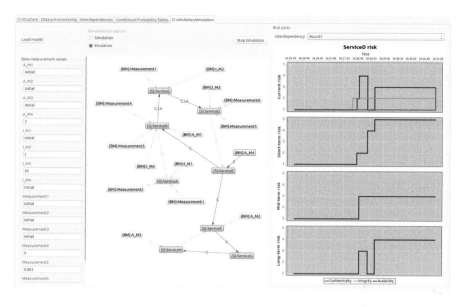

Fig. 12. Screen-shot of risk simulation/emulation tab

the Java SWT (Standard Widget Toolkit) (http://www.eclipse.org/swt/) for all the GUI components.

The tool was implemented using Ubuntu 12.10 operating system. Although the tool was implemented using Java, which should provide multi-platform support, the tool was only tested in Linux and therefore multi-platform support is not guaranteed.

5   Conclusions and Future Work

In this article a tool based on a novel CI risk modelling approach for CI risk monitoring based on Bayesian networks was presented. The idea of Bayesian network based CI risk modelling is to represent CIs as services they provide and the dependencies between the services. Risk is estimated based on observing system measurements that represent the CI service state as well as on observing risk of other CI services the service depends on. This model is represented as a BN where the nodes represent CI services and system measurements, and the edges represent dependencies between nodes. The probabilities in the CPT represent the most likely CI service risk, considering the state of the parent nodes (system measurements or CI services). Those probabilities can be learned from data records as well as being assigned by experts based on their experience with CI operation. Using a Bayesian approach for the CI security model allows some advanced features like risk prediction or handling of interdependencies.

The tool presented in this work fully implements the idea of Bayesian network based CI security modelling. Therefore, the tool is divided into five separate parts, realized in the GUI as five tabs of a tab folder: The *CI structure* tab helps domain experts in generating the structure of the CI security model by providing a simple visual representation. The *data pre-processing* tab allows to pre-process the dataset used to learn the conditional probabilities and complement it with data that needs to be provided by domain experts. The *interdependencies* tab allows to identify interdependencies (directed cycles in the graph model) and supplement them with data needed to learn conditional probabilities when interdependencies are present. The *conditional probabilities* tab allows to learn CI service CPTs for current risk as well as risk prediction. It also supports special handling of CI services that are part of an interdependency. Finally, the *CI simulation/emulation* tab allows to evaluate service risk propagation by simulating or emulating base measurement states. The presented tool is available as open source software. For more information please contact the authors.

Future work will focus on evaluation of the Bayesian network based CI risk model as well as the tool presented in this work in the context of a case study. The first part of the evaluation process will investigate the feasibility of CI analysis. The main question in this phase is if it is possible to extract critical CI services, system measurements and dependencies from complex systems and if the graphical representation of the model using the tool can help to simplify the task. The second part of evaluation concerns CPT learning. The main questions in this phase will be if the BN approach will be manageable or if there will be too many evidence variables and therefore too complex CPTs, if the probabilities can be sufficiently learned from data and if expert estimation of probabilities is feasible. Furthermore, one interesting aspect of evaluation will be to see if the implemented learning algorithms of the tool are fast enough to handle CPT learning using real-world datasets. In the last phase of the validation the usefulness of a CI risk monitor will be evaluated with CI operators, using the presented simulation/emulation approach of the tool as a basis for evaluation.

Acknowledgements. One of the authors would like to thank the Luxembourgish National Research Fund (FNR) for funding his PhD work under AFR grant number PHD-09-103. Also, the authors would like to thank the developers of the open-source libraries mentioned in Section 4.6 for providing the technologies used to implement this tool. The authors thank the Grid'5000 project for cooperation and for providing realistic data sets.

References

1. Aubert, J., Schaberreiter, T., Incoul, C., Khadraoui, D.: Real-time security monitoring of interdependent services in critical infrastructures. case study of a risk-based approach. In: 21th European Safety and Reliability Conference, ESREL 2010 (September 2010)
2. Aubert, J., Schaberreiter, T., Incoul, C., Khadraoui, D., Gateau, B.: Risk-based methodology for real-time security monitoring of interdependent services in critical infrastructures. In: ARES 2010 International Conference on Availability, Reliability, and Security, pp. 262–267 (February 2010)

3. Baiardi, F., Telmon, C., Sgandurra, D.: Hierarchical, Model-based Risk Management of Critical Infrastructures. Reliability Engineering & System Safety (9), 1403–1415 (2009); ESREL 2007, the 18th European Safety and Reliability Conference
4. Caldeira, F., Schaberreiter, T., Monteiro, E., Aubert, J., Simoes, P., Khadraoui, D.: Trust based interdependency weighting for on-line risk monitoring in interdependent critical infrastructures. In: 2011 6th International Conference on Risk and Security of Internet and Systems (CRiSIS), pp. 1–7 (September 2011)
5. Eronen, J., Laakso, M.: A case for protocol dependency. In: IEEE International Workshop on Critical Infrastructure Protection, pp. 22–32 (2005)
6. Haimes, Y.Y., Kaplan, S., Lambert, J.H.: Risk filtering, ranking, and management framework using hierarchical holographic modeling. In: Risk Analysis. Society for Risk Analysis (2002)
7. Haslum, K., Arnes, A.: Multisensor real-time risk assessment using continuous-time hidden markov models. In: 2006 International Conference on Computational Intelligence and Security, pp. 1536–1540 (2006)
8. Panzieri, S., Setola, R., Ulivi, G.: An approach to model complex interdependent infrastructures. In: 16th IFAC World Congress, CISIA, Critical Infrastructures (2005)
9. Pietikäinen, P., Karjalainen, K., Eronen, J., Röning, J.: Socio-technical security assessment of a voip system. In: The Fourth International Conference on Emerging Security Information, Systems and Technologies, SECURWARE 2010 (July 2010)
10. Rinaldi, S.M., Peerenboom, J.P., Kelly, T.K.: Identifying, understanding, and analyzing critical infrastructure interdependencies. IEEE Control Systems Magazine, 11–25 (2001)
11. Schaberreiter, T., Aubert, J., Khadraoui, D.: Critical infrastructure security modelling and resci-monitor: A risk based critical infrastructure model. In: IST-Africa Conference Proceedings, pp. 1–8 (May 2011)
12. Schaberreiter, T., Bonhomme, C., Aubert, J., Incoul, C., Khadraoui, D.: Support tool development for real-time risk prediction in interdependent critical infrastructures. In: Risk and Trust in Extended Enterprises (RTEE 2010) Workshop. ISSRE Wksp 2010. IEEE International Symposium on Sofware Reliability Engineering (November 2010)
13. Schaberreiter, T., Bouvry, P., Röning, J., Khadraoui, D.: A bayesian network based critical infrastructure risk model. In: Schütze, O., Coello Coello, C.A., Tantar, A.-A., Tantar, E., Bouvry, P., Del Moral, P., Legrand, P. (eds.) EVOLVE - A Bridge Between Probability, Set Oriented Numerics, and Evolutionary Computation II. AISC, vol. 175, pp. 207–218. Springer, Heidelberg (2012)
14. Schaberreiter, T., Caldeira, F., Aubert, J., Monteiro, E., Khadraoui, D., Simones, P.: Assurance and trust indicators to evaluate accuracy of on-line risk in critical infrastructures. In: 6th International Conference on Critical Information Infrastructure Security, CRITIS 2011 (2011)
15. Schaberreiter, T., Kittilä, K., Halunen, K., Röning, J., Khadraoui, D.: Risk assessment in critical infrastructure security modelling based on dependency analysis (short paper). In: 6th International Conference on Critical Information Infrastructure Security, CRITIS 2011 (2011)
16. Sokolowski, J., Turnitsa, C., Diallo, S.: A conceptual modeling method for critical infrastructure modeling. In: 41st Annual Simulation Symposium, ANSS 2008, pp. 203–211 (April 2008)

Challenges on Probabilistic Modeling for Evolving Networks

Jianguo Ding[1] and Pascal Bouvry[2]

[1] Interdisciplinary Center for Security, Reliability and Trust
University of Luxembourg, L-1359 Luxembourg
Jianguo.Ding@ieee.org
[2] Faculty of Science, Technology and Communication
University of Luxembourg, L-1359 Luxembourg
Pascal.Bouvry@uni.lu

Abstract. With the emerging of new networks, such as wireless sensor networks, vehicle networks, P2P networks, cloud computing, mobile Internet, or social networks, the network dynamics and complexity expands from system design, hardware, software, protocols, structures, integration, evolution, application, even to business goals. Thus the dynamics and uncertainty are unavoidable characteristics, which come from the regular network evolution and unexpected hardware defects, unavoidable software errors, incomplete management information and dependency relationship between the entities among the emerging complex networks. Due to the complexity of emerging networks, it is not always possible to build precise models in modeling and optimization (local and global) for networks. This paper presents a survey on probabilistic modeling for evolving networks and identifies the new challenges which emerge on the probabilistic models and optimization strategies in the potential application areas of network performance, network management and network security for evolving networks.

Keywords: network evolution, probabilistic modeling, dynamics and uncertainty.

1 Introduction

It is recognized that three laws, Moore's law, Gilder's law and Metcalfe's law, which are governing the spread of technology and are related to the rapid evolution of IT networks [10]. Moore's law indicates the computing capability of computers doubles every 18 months. Gilder's law claims the total bandwidth of communication systems triples every 12 months for the next 25 years. Metcalfe's law presents the value of a telecommunications networks is proportional to the square of the number of connected users of the systems (n^2).

The typical evolution in networks is paralleled with following changes: the improved/degraded hardware performance, updated software (system software, application software) and its functions, extended network structure with the integration of emerging heterogeneous networks (mobile communication networks,

O. Schütze et al. (eds.), *EVOLVE - A Bridge between Probability, Set Oriented Numerics, and Evolutionary Computation III,* Studies in Computational Intelligence 500,
DOI: 10.1007/978-3-319-01460-9_4, © Springer International Publishing Switzerland 2014

sensor networks, ad hoc networks, vehicle networks, overlay networks, and Internet of Things), extension of network scale with increasing wired/mobile and wireless devices joined, updated network protocols, improved network functions (from information exchanging to complex online transaction and numerous services), dynamic functional evolution, variation of the network performance, and emerging network applications and network services.

The evolution of networks is not only on physical networks, but also on information networks and service networks, which is over physical networks. In real IT application scenarios, the evolution is characterized with the combinational evolving results on physical networks, information networks and service networks.

The evolving networks demonstrate the complex changes in network structure, network functions, network performance, and interoperation relationship with the time evolving, and thus more opportunistic networks and self-organizing networks come into being one trend of the network evolution. The emerging computing models (distributed computing, pervasive computing, cognitive computing, opportunistic computing, scalable computing, autonomic computing, physical computing, and probabilistic computing), which are employed to model and manage the complex dynamic networks and pertain to the operation, administration, maintenance, and provision of networked systems for secure (reliable) and effective network performance [11].

The dynamics and uncertainty are unavoidable characteristics, which come from the regular network evolution and unexpected hardware defects, unavoidable software errors, incomplete management information and dependency relationship between the entities among the emerging complex networks. Due to the complexity of emerging networks, it is not always possible to build precise models in modeling and optimization (local and global) for networks. New challenges emerge on the probabilistic models and optimization strategies in the areas of network performance, network management, network security for evolving IT networks.

This paper presents a systematic survey on the probabilistic modeling for evolving networks and identifies the new challenges which emerge on the probabilistic models and optimization strategies in the potential application areas of network performance, network management and network security for evolving networks.

2   Emerging Characteristics in Evolving Networks

The evolving IT networks demonstrate emerging characteristics as follows:

1. Dynamics
   - Some networks are running in dynamic style by nature, such as mobile communication networks, wireless sensor networks, vehicle networks, and overlay networks (P2P, VPN). These networks can be organized as opportunistic networks or self-organizing networks. That means the structure of the networks is changing over the topology, with the variation on routers, mobile servers and mobile clients.

- The performance (robustness) of individual network components, such as routers, servers, clients or other key network devices/services vary with the network evolution. Some components are improved or degraded with the hardware performance, technical improvement, or systematic evolution.
- The performance (robustness) of the individual links, which demonstrate the dependencies between main network components, changes with the structural or functional modification of the networks.
- Both local and global changes are interdependent. That means any local changes may result in the variation of global network performance. Any global modification can result in the changes in local network performance as well.
- Theoretically, the network function and structure have strong interdependence [22]. The evolving structure of networks will bring the changes in network functions. On the other side, it is possible that network function modification can result in the redesign/reconfiguration of network structure.

2. Heterogeneity
  In IT networks, there are 2 types of heterogeneous networks, integrated networks and overlay networks.
    - Integrated networks
      With the advances of emerging networks, more heterogeneous networks are integrated. For example, the sensor networks are integrated with local networks, vehicle networks join mobile communication networks, heterogeneous network devices are integrated into Internet, and the trend of Internet of Things emerges. The heterogeneous network integration demonstrates the integration of different structures, different functions, different performance, different network protocols, different software components, and even different services. Integrated networks are not only the accumulation of networks, but also updated properties and functions merging with the evolution.
    - Overlay networks
      An overlay network is a computer network which is built on the top of another network. Nodes in the overlay networks can be connected by virtual or logical links, each of which corresponds to a path, perhaps through many physical links, in the underlying networks. Information networks and service networks based on the physical networks are commonly identified as overlay networks as well. Overlay networks organize peers with different strategies, thus their topology and routing performance are different. The consequent reliability and fault resiliency varies as well [16]. Overlay networks are organized by spontaneous and dynamic connectivity between users/clients, this evolution model is accompanied with the continuing structure dynamics.

3. Temporal Networks
  Evolving is a time correlated process. The evolving network structure, describing how the network is wired and how the abstract nodes are connected,

helps us to understand, predict and optimize the behaviour of dynamic networks. In many cases, however, the edges/links are not continuously active. In some cases, edges/links are active for non-negligible periods of time. Like network topology, the temporal structure of edge activations can affect dynamics of systems interaction through the network. The dynamic weights, which indicate the interdependencies between networks components, demonstrate the network evolving process as well. An evolving networks is a typical temporal network, which can be modelled to elucidate the behaviour of a dynamic system. The fundamental properties in temporal networks are quite different from those for static networks.

4. Complexity
   Complexity has an important relationship to resilience and the robustness of systems, because resilience mechanisms such as self-organization and autonomic behaviour increase complexity, and increased complexity may result in greater network vulnerability [24]. The complexity in evolving IT networks comes from structural complexity, network evolution, connection diversity, dynamical complexity, nodes diversity, meta-complication. Furthermore, the various complications can influence each other.

5. Macro view vs. micro view
   In evolving complex networks, self-organizing processes are deployed at multiple levels. Challenging questions about the dynamics of micro-macro transition include: (i) how are emergent properties related to micro interactions? (ii) how can we reverse-engineer the mechanics of complex system from their behaviour under a controlled set of external stimuli? Thus the interrelationship between micro and macro behaviour in evolving networks is rather important in application scenarios.

6. Probability
   In evolving network, the system dynamics and the intrinsic complexity make the complex networks with probabilistic properties. The incomplete and uncertain information need to be integrated into the research models, so that the system models can be more reasonable and realistic [9]. Thus the time based probabilistic factors should be embedded into the network modeling in evolving networks.

3   Challenges for Probabilistic Modeling

1. Structure dynamics
   In IT networks, the structure evolving is paralleled with the changes on:

   - Network nodes: New nodes (network hardware/software components) are added or removed from the network, or the improvement/degradation in the performance of nodes.
   - Network links: New links (interoperation/interdependency) are added or removed from the network, or the improvement/degradation in the performance of links.

- Weight of the dependencies: The weight of the dependencies between network components indicates the measurement of the importance on the performance or dependencies among the related components. It is a time related function during the evolution.

Two types of dynamics are required in order to fully understand how networks evolve over time. The type of dynamics most commonly used in IT network analysis are those dealing with the nature of interactions between nodes as a consequence of the network structure, which is called **dynamics on the network**. This category governs how nodes react to each other based on the overall structure of the network. The second type of network dynamics, named **dynamics of the network**, governs the changes in the network structure and evolution of the structure.

There are three approaches to model the dynamic networks:

- The evolution of a network can be described as a sequence of static networks and since there exist many parameters to describe accurately a static network, one can study the evolution of the network through the evolution of these parameters.
- The evolution itself can be studied with defined parameters to capture the evolving properties, such as the rate of appearance or disappearance of nodes and edges.
- An intermediate approach can be used which consists specific phenomena in studying or users of interest with time.

The approach selection is based on the specific scenario. For example, if the network structure keeps stable and with minor modification, the sequence snapshot of the static networks can be considered as an appropriate evolving model. The network infrastructure and network backbone follow this class. But suppose the network structure evolves with great changes, and then the parameter modelling or intermediate approach might be appropriate. Mobile communication networks, sensor networks, vehicle networks belong to this class.

In modelling the evolving networks, not only the structure properties (topologies, nodes, and links) are included, but also the non-structural properties (weights, importance, functions) should be considered. It is also a challenge to model the dynamic structure and evolving properties in integrated heterogeneous network and overlay networks.

2. Stochastic dynamics

Along the network evolution, some changes will generate new data characteristics which might not to be the property of the historical data. Generally, the building process of the network and the parameter estimation requires more data as the number of variables varies, as long as the accuracy in the estimations and in the network topology is to be maintained. However, the network evolution is not necessarily to maintain the accuracy in the parameter estimation because of the unstable and dynamic properties. There is some difficulties in obtaining a stable historical data for an evolving network on application scenarios.

3. Model of heterogeneous networks
   Evolving networks are composed of heterogeneous networks in network structures and functions. This will make the integrated network model more complex, since some unbalanced and heterogenous network sections are not integrated consistently in following unique or common principles in the structure and the functions.

4. Relationship between macro and micro networks
   For a large scale heterogeneous network, the macro characteristics are strongly related to the micro pieces of the network and vice versa. When modeling the global probabilistic network, the local subnets modeling and their merging style between pieces are rather important, in which the interdependencies can be identified and measured reasonably. However, large scale complex networks demonstrate new properties which is hard to be identified from micro networks, such as small world property, scale free network, etc.

5. Control and feedback
   In modeling network application and services, the control and feedback loop is inevitable at different levels, particularly in logical and service networks. The loops among networks are apt to make the related modeling out of control. The overlap (repeat) dependencies vague the relationship between the networked components.

6. Computing complexity
   The complexity in emerging networks makes medium size models usually intractable, since the number of variables involved is greater than in static models. Highly connected networks and dynamic changes among the network structure and dependencies between related components make the evolving networks total complex dynamic systems, and thus brings very challenging problems in computing complexity.

7. Probabilistic factors
   The dynamic and complex network behaviours inevitable brings probabilistic factors to the evolving networks. Thus probabilistic factors should be included in the models of evolving networks. The combination of probabilistic models and complex network model challenges the modelling of dynamic evolving networks.

4   Probabilistic Modeling for Dynamic Networks

The goal of modelling for evolving networks is to model the state of a system and its evolution over time in a richer and more natural way. It is widely recognized that probabilistic graphical models provide a good framework for both knowledge representation and probabilistic inference for dynamic evolving networks.

A probabilistic dynamic model will be considered as a sequence of graphs indexed by the time, representing the temporal evolution of a system. Each graph symbolizes the state of the system and the dependencies among its components at a given time. The dynamic behaviour of the components of the system is described by a set of temporal dependencies among these components in different time slices. Furthermore, these dependencies are quantified by conditional probability tables associated with the components of the system. In order to make the management of such models feasible, a set of restrictions must be considered for both its qualitative and quantitative aspects.

4.1 Dynamic Graphs

The dynamic behaviour of any specific system which changes over time requires an implicit or explicit time representation. To model such systems is a very important task: the initial structure of the model and its propagation over time, the probabilities attached to the structure, the qualitative and quantitative interrelations among variables in different time slices, etc., need to be taken into account [18].

Basically, a network can be modelled as a graph, which includes essential elements: nodes, links, and weighs on links or/and nodes.

A graph can be defined as a triple (V, E, f_V, f_E) where V is a set of vertexes, E is a set of edges u, v, and f is a function, $f_V : V \to N$, $f_E : E \to N$, where N is some number system, assigning a value or a weight. Depending on the context, the weights may be real numbers, complex numbers, integers, elements of some group, etc. A network or fully weighted graph has weights assigned to both nodes and edges.

These definitions of (static) graphs and networks involve the following entities: V (a set of nodes), E (a set of edges), f_V (mapping vertexes to numbers), f_E (mapping edges to numbers). A dynamic graph is obtained when any of these four entities changes over time. Thus, there are several basic kinds of dynamic graphs.

- in a node-dynamic graph or digraph, the set V varies with time. Thus, some nodes may be added or removed. When nodes are removed, the edges incident with them are also eliminated.
- in an edge-dynamic (or arc-dynamic) graph or digraph, the set E varies with time. Thus, edges may be added or deleted from the graph or digraph.
- in a node weighted dynamic graph, the function f_V varies with time; thus, the weights on the nodes also change.
- in an edge weighted dynamic graph or digraph, the function f_E varies with time.
- in fully weighted dynamic graph, both functions f_V and f_E may vary with time.

Thus a dynamic graph is defined as a triple with time parameter t: (V^t, E^t, f_V^t, f_E^t). Harary classifies the dynamic graphs by the change of any of these [14]:

1. Node dynamic graphs where the vertex set V^t changes over time t.
2. Edge dynamic graphs where the edge set E^t change over time t.
3. Node weighted dynamic graphs where the f_V^t function varies with time t.
4. Edge weighted dynamic graphs where the f_E^t function varies with time t.

All combinations of the above types can occur. For example, a computer network with changing bandwidth (edge-weight), changing topology (edges being added or deleted), changing computing power (node-weights changing), and computers representing nodes crashing, and recovering represents a dynamic graph that entails all the above basic types.

Work on dynamic graph theory have been motivated by finding patterns and laws. Power laws, small diameters, shrinking diameters have been observed. Graph generation models that try to capture these properties are proposed to synthetically generate such networks [5]. There are several problems to be answered in these complex dynamic networks.

- Is the network evolving normally?
- What is normal behaviour of the network?
- Is there a phase transition along the network evolving?

There is a strong correlation between finding patterns in static graphs and dynamic evolving graphs.

Graph similarity functions, which is used to measure the degree of the dynamics on networks, are categorized into two groups:

- feature based similarity measures
- structure based similarity measures

Using the topology of the graphs, two similarity metrics have been defined, maximum common subgraph distance and the graph edit distance.

Graph clustering has become a central tool for the analysis of dynamic networks in general, with applications ranging from the field of social sciences to biology and to the growing field of complex systems. The general aim of graph clustering is to identify dense subgraphs in networks. Countless formalizations thereof exist, however, the overwhelming majority of algorithms for graph clustering relies on heuristics, e.g., for some NP-hard optimization problem, and do not allow for any structural guarantee on their output.

4.2   Power Law Random Graphs

Random graphs can date back to the work of Erdős and Rényi for the theory of random graphs [12]. The random graph model $G(n, e)$ assigns uniform probability to all graphs with n nodes and e edges while in the random graph model $\mathcal{G}(n, p)$ each edge is chosen with probability p.

Power law random graph model [2] is an extension of random graph, whose degree distribution follows a power law. Most of IT network system have this properties. Power law rand graph model has two parameters. The two parameters only roughly delineate the size and density but they are natural and convenient for describing a power law degree sequence. The power law random graph model $P(\alpha, \beta)$ is described as follows. Let y be the number of nodes with degree x. $P(\alpha, \beta)$ assigns uniform probability to all graphs with $y = e^{\alpha}/x^{\beta}$ (where self loops are allowed). Note that α is the intercept and β is the (negative) slope when the degree sequence is plotted on a log-log scale.

There is also an alternative power law random graph model analogous to the uniform graph model $\mathcal{G}(n, p)$. Instead of having a fixed degree sequence, the random graph has an expected degree sequence distribution. The two models are basically asymptotically equivalent, subject to bounding error estimates of the variances.

The power law random graph model provides an approach to model the dynamic evolving complex networks. However, there are some questions that remain to be resolved. For example, what is the effect of time scaling? How does it correspond with the evolution of β? What are the structural behaviours of the power law random graphs?

4.3   Dynamic Flowgraph Methodology

The dynamic flowgraph methodology (DFM) [6] is an approach to model and analyze the behaviour of dynamic systems for reliability/safety assessment and verification. DFM models express the logic of the system in terms of causal relationships between physical variables and states of the system. The time aspects of the system (execution of control commands, dynamics of the process) are represented as a series of discrete state transitions. DFM can be used for identifying how certain postulated events may occur in a system. The result is a set of timed fault trees, whose prime implicants (multi-state analogue of minimal cut sets) can be used to identify system faults resulting from unanticipated combinations of software logic errors, hardware failures, human errors and adverse environmental conditions.

DFM models are directed graphs, analyzed by discrete time instances. They consist of variable and condition nodes; causality and condition edges; and transfer and transition boxes and their associated decision tables. A node represents a variable that can be in one of a finite number of predefined states. The state of a node can change at discrete time instances. The state of the node is determined by the states of its input nodes. Each node can have several inputs but only one output (its state). The state of the node can act as an input to possibly several other nodes. The state of a node at time t is determined by

- the states of its input nodes at a single instance of time (say, $t - n$),
- the lag n, an integer that tells how many time instances it takes for an input to cause the state of the present node.

The state of a node, as a function of the states of its input nodes, is determined by a decision table. A decision table is an extension of the truth table where each variable can be represented with any finite number of states. The decision table contains a row for each possible combination of input variable states. The maximum possible number of rows in the decision table is the product of the numbers of states of the input nodes.

After construction, the DFM model can be analyzed in two different modes, deductive and inductive. In inductive analysis, event sequences are traced from causes to effects; this corresponds to simulation of the model. In deductive analysis, event sequences are traced backward from effects to causes.

A deductive analysis starts with the identification of a particular system condition of interest (a top event); usually this condition corresponds to a failure. To find the root causes of the top event, the model is backtracked for a predefined number of steps through the network of nodes, edges, and boxes. This means that the model is worked backward in the cause-and-effect flow to find what states of variables (and at what time instances) are needed to produce the top event. The result of a deductive analysis is a set of prime implicants.

A prime implicant consists of a set of triplets (V, S, T); each triplet tells that variable V is in a state S at time T. The circumstances described by the set of triplets cause the top event. Prime implicants are similar to minimal cut sets of fault tree analysis, except that prime implicants are timed and they deal with multi-valued variables (fault trees deal with Boolean variables). A useful analogy is that deductive analysis corresponds to minimal cut set search of a fault tree. Once primary implicants have been found, the top event probability can be quantified in a fault tree.

For large scale dynamic networks, the state analysis and the fault propagation among dynamic networks make the resolution intangible with huge computing complexity.

4.4 Dynamic Factor Graphs

Directed and undirected networks coexist in most IT networks. For large networks (graphs), the factorization properties of a graphical model, whether undirected or directed, may be difficult to visualize from the usual depictions of graphs. The formalism of factor graphs provides an alternative graphical representation, one which emphasizes the factorization of the distribution.

Let F represent an index set for the set of factors defining a graphical model distribution. In the undirected case, this set indexes the collection C of cliques, while in the directed case F indexes the set of parent-child neighborhoods. We then consider a bipartite graph $G = (V, E, F)$, where V is the original set of vertexes, and E is a new edge set, joining only vertexes $s \in V$ to factors $a \in F$. In particular, edge $(s, a) \in E$ if and only if x_s participates in the factor indexed by $a \in F$.

For undirected models, the factor graph representation is of particular value when C consists of more than the maximal cliques. Indeed, the compatibility functions for the nonmaximal cliques do not have an explicit representation in the usual representation of an undirected graph. However, the factor graph makes them explicit.

Time series collected from real-world phenomena are often an incomplete picture of a complex underlying dynamical process with a high-dimensional state that cannot be directly observed.

The simplest approach to modeling time series relies on time-delay embedding: the model learns to predict one sample from a number of past samples with a limited temporal period. This method can use linear auto-regressive models, as well as non-linear ones based on kernel methods (e.g. support-vector regression), neural networks (including convolutional networks such as time delay neural networks), and other non-linear regression models. Unfortunately, these approaches have difficult in capturing hidden dynamics with long-term dependency because the state information is only accessible indirectly through a (possibly very long) sequence of observations [7].

To capture long-term dynamical dependencies, the model must have an internal state with dynamical constraints that predict the state at a given time from the states and observations at previous times (e.g. a state-space model). In general, the dependencies between state and observation variables can be expressed in the form of a Factor Graph for sequential data, in which a graph motif is replicated at every time step.

For a complex and non-linear system, a model might allow the use of complex functions to predict the state and observations, and will sacrifice the probabilistic nature of the inference. Instead, the inference process (including during learning) will produce the most likely (minimum energy) sequence of states given the observations. Dynamic Factor Graph (DFG) is a natural extension of Factor Graphs specifically tuned for sequential data. To model complex dynamics, DFG allows the state at a given time to depend on the states and observations over several past time steps.

Dynamical Factor Graphs manage to perfectly reconstruct multiple oscillatory sources or a multivariate chaotic attractor from an observed one-dimensional time series. DFGs also outperform Kalman Smoothers and other neural network techniques on a chaotic time series prediction tasks, DFGs can be used for the estimation of missing motion capture data. Proper regularization such as smoothness or a sparsity penalty on the parameters enable to avoid trivial solutions for high-dimensional latent variables [19].

4.5   Time-Varying Graphs

Time-varying graphs (TVG) have been a topic of active research recently in the study of communication networks with intermittent connectivity such as delay-tolerant networks and even disruption-tolerant social networks; duty cycling wireless sensor networks, and so on [8]. Existing research on time-varying graphs ranges from algorithmic studies on graph journeys to analysis of specific properties such as flooding time in dynamic random graphs. Empirical simulation-based analysis of certain temporal graph properties such as temporal distance and temporal efficiency are hot topics in this area.

The TVG can describe a multitude of different scenarios, from transportation networks to communication networks, complex systems, or social networks. Some research questions are generated by the application requirements in dynamic networks.

One important task is to explore the universe of dynamic networks using the formal tools provided by the TVG formalism. The long-term goal is to provide a comprehensive map of this universe, to identify both the commonality and the natural differences between the various types of dynamical systems modeled by TVG.

The design and analysis of distributed algorithms and protocols for time-varying graphs is an open research area. In fact very few problems have been attacked so far: routing and broadcasting in delay-tolerant networks; broadcasting and exploration in opportunistic-mobility networks; new self-stabilization techniques; detection of emergence and resilience of communities, and viral marketing in social networks.

If the interactions in a network can be planned and decided by a designer, then a number of new interesting optimization problems arise with the design of time-varying graph. They may concern, for example, the minimization of the temporal diameter or the balancing of nodes eccentricities.

Analyzing the complexity of a distributed algorithm in a TVG , e.g. in number of messages, is not trivial, partly because contrarily to the static cases, the complexity of an algorithm in a dynamic network has a strong dependency, not only on the usual network parameters (number of nodes, edges, etc.), but also on the number of topological events taking place during its execution. In many of the algorithms, the majority of messages is in fact directly triggered by topological events, e.g., in reaction to the local appearance or disappearance of an edge. The number of topological events therefore represents a new complexity parameter, whose impact on various problems remains to study.

Through the use of the interaction-centric point view, TVGs enable to look at the interplay between topological aspects that allow local interaction to have global effects.

4.6   Dynamic Bayesian Networks

Dynamic Beyesian Networks (DBN), which is an extension of causal probabilistic networks [4] and static Bayesian networks, is to model a system that is dynamically changing or evolving over time. This model will enable users to monitor and update the system as time proceeds, and even predict further behaviour of the system. In every time slice of a temporal model corresponds to one particular state of a system, and if the movement between the slices reflects a change in state instead of time.

Dynamic Beyesian Networks are usually defined as special case of singly connected Bayesian networks specifically aimed at time series modelling.

All the nodes, edges and probabilities that form static interpretation of a system is identical to a Bayesian network. Variables can be denoted as the sate of a DBN, because they include a temporal dimension. The states of any system described as a DBN satisfy the Markovian condition, that is defined as follows: The sate of a system at time t depends only on its immediate past, i.e. its state at time $t-1$. Also, this property is frequently considered as a definition of First order Markov property: the future is independent of the past given the present. The states of a dynamic model do not need to be directly observable. They may influence some other variables that we can directly measure or calculate. Also, the state of some system needs not to be a unique, simple state. It may be regarded as a complex structure of interacting states. Each state in a dynamic model at one time instance may depend on one or more states at the previous time instance or/and on some states in the same time instance [21]. So, in DBN states of a system at time t may depend on system states at time $t - 1$ and possibly on current states of some other nodes in the fragment of DBN structure that represents variables at time t.

It is not easy to model time and uncertainty in a way that clearly and adequately represents the problem domains at hand. Related approaches can be classified into three broad categories:

- Models that use static BNs and formal grammars to represent temporal dimension (known as Probabilistic Temporal Networks)
- Models that use mixture of probabilistic and non-probabilistic frameworks
- Models that introduce temporal nodes into static BNs structure to represent time dependence.

By using a DBN, we assume that dynamic data are generated sequentially by some hidden states of a dynamic factor evolving over time. Since the hidden states cannot be observed directly, they can only be inferred from the observed data given a learned DBN. Learning a DBN involves estimating both its structure and parameters from data [25]. The structure of a DBN refers primarily to (1) the number of hidden states of each hidden variable in a model and (2) the conditional dependence among hidden states of all the hidden variables of a model, i.e, factorization of the model state space for determining the topology of a graph network. There have been extensive studies in the machine learning community on efficient parameter learning when the structure of a model is known a priori. Mixed atemporal and temporal independence relations among DBN models is examined as well [15]. However, much less efforts have been made to tackle the more challenging problem of learning the optimal structure of an unknown DBN. As a consequence, most previous DBN-based data modelling approaches avoid the structure learning problem by setting the structure manually. However, it has been shown that a learned structure can be advantageous over those that are manually set [13].

DBNs represent the state of the world as a set of variables, and model the probabilistic dependencies of the variables within and between time steps. While a major advance over previous approaches, DBNs are still unable to compactly represent many real-world domains. In particular, domains can contain multiple objects and classes of objects, as well as multiple kinds of relations among them; and objects and relations can appear and disappear over time. Capturing such a domain in a DBN would require exhaustively representing all possible objects and relations among them. This raises two problems. The first one is that the computational cost of using such a DBN would likely be prohibitive. The second one is that reducing the rich structure of the domain to a very large "flat" DBN would render it essentially incomprehensible to human beings [20].

4.7 Probabilistic Complex Networks

The term of "complex network" is usually used for referring the natural networks that are usually complex and cannot be modeled just through random graphs. Most real-world networks have these complex topological features, such as, heavy-tail in the degree distribution, high clustering coefficient, assortativity or disassortativity among vertexes, inherent multiparty structure, self-similar hierarchical structure, etc. Clustering coefficient represents the ratio of a network that satisfies your friends are also mutually friends. The assortativity represents the grouping of nodes. Inherent multiparty structure indicates that there exists many intrinsic multiparty properties in the real world.

On the contrary, simple networks usually have these properties. For instance, they can be represented by graphs such as a lattice or random graph. The topological structure is roughly the same in any part of network. And, they do not posses the above complex network features. Examples of complex networks include social networks, computer networks, biological networks - neurons, or protein structure, river networks, power-line networks, etc.

A probabilistic complex network can be defined as a set of probabilistic nodes and probabilistic edges in a network topology which follows the characteristics of complex networks, such as power-law of the degree distribution of nodes and the small world phenomenon which specifies the shortest path between any two nodes are generally small. The network topology can be represented by setting the probabilities to zero in some of the edges at a completely connected network. A probabilistic node means there are possibly either discrete states or continuous value (or attribute vector) in the node. Similarly, a probabilistic edge represents that the discrete state or value of edges is probabilistic [17].

In a probabilistic complex network, there should be causal relationships between the states/values of nodes and edges. This is the biggest difference between a probabilistic complex network and a random graph. The inter-nodes, inter-edges, and nodes-edges state relationships can be determined by a deterministic or probabilistic model by some pre-specified rules. When defining dynamic probabilistic complex network as a probabilistic complex network, evolves over time will be considered. The probably distributions of nodes and edges can be different at any sampled time. For simplicity, in IT networks, we only consider the discrete cases for probabilistic complex networks.

Many real-life network systems (such as Internet, WWW, etc.) can be modeled as probabilistic complex networks of interacting components. Although the study of such large scale networks is not new, there has recently been much renewed interest in this field. This is due to technological advances of two types: (i) the collection of data which depict large networks in detail, and (ii) the development of computational tools for the analysis of data. Among the well-studied examples of such networks are the World Wide Web, citation networks, neuronal connections, metabolic networks, ecological webs and more [1] .

Traditional Erdős and Rényi random graph models have possion degree distributions. However, it has been found that many real life networks follow power law distributions. Generalized random graph models have been proposed to mimic the power law degree distribution of the real networks but these models do not explain how such a phenomena occurs in these graphs. Barabasi et al.[3] introduced a model (BA Model) the concept of preferential attachment for this purpose.

The BA model is an algorithm for generating random scale-free networks using a preferential attachment mechanism. Scale-free networks are widely observed in natural and human-made systems, including the Internet, the world wide web, citation networks, and some social networks.

The network begins with an initial network of m_0 nodes ($m_0 \geq 2$) and the degree of each node in the initial network should be at least 1, otherwise it will always remain disconnected from the rest of the network. New nodes are added to the network one at a time. Each new node is connected to existing nodes with a probability that is proportional to the number of links that the existing nodes already have. Formally, the probability p_i that the new node is connected to node i is

$$p_i = \frac{k_i}{\Sigma_j k_j} \tag{1}$$

where k_i is the degree of node i and the sum is made over all preexisting nodes j. Heavily linked nodes ("hubs") tend to quickly accumulate even more links, while nodes with only a few links are unlikely to be chosen as the destination for a new link. The new nodes have a "preference" to attach themselves to the already heavily linked nodes.

Follow the BA model, there are two types of sub-models:

- Model A retains growth but does not include preferential attachment. The probability of a new node connecting to any pre-existing node is equal. The resulting degree distribution in the limit is geometric [23], indicating that growth alone is not sufficient to produce a scale-free structure.
- Model B retains preferential attachment but eliminates growth. The model begins with a fixed number of disconnected nodes and adds links, preferentially choosing high degree nodes as link destinations. Though the degree distribution early in the simulation looks scale-free, the distribution is not stable, and it eventually becomes nearly Gaussian as the network nears saturation. So preferential attachment alone is not sufficient to produce a scale-free structure. The failure of models A and B to lead to a scale-free distribution indicates that growth and preferential attachment are needed simultaneously to reproduce the stationary power-law distribution observed in real networks [1].

In modelling probabilistic complex networks, still some challenges forward:

1. Finding dynamic community structures in a probabilistic complex networks,
2. Investigating appropriate classification techniques,
3. Applying on various practical datasets,
4. Processing large datasets,
5. Investigating the probabilistic approach to do information classification and network topology learning.

5   Conclusions

With the evolution in network structures and the applications, more requirements are generated for the modelling of probabilistic dynamic networks. The
challenges include structure dynamics, probabilistic factors, heterogeneous networks, control and feedback, computing complexity, etc. There are several
approaches in modelling the dynamic and probabilistic factors and holistic evolving networks, such as dynamic graph, power law random graph, dynamic flowgraph method, dynamic factor graphs modeling, time-varying graphs, dynamic
Bayesian Networks, probabilistic complex network modelling. However, most of
the approaches are linked to application scenarios and focus on specific dynamic
systems and on particular dynamic behaviours. There is no common approach
available to deal with the various dynamics among the evolving networks.

Based on the observation and requirements on empirical research, the following topics are deserved for detailed investigation:

- dynamic properties of the network should be identified. This includes the
  network structure changes, the scale, degree and speed of the changes among
  networks;
- the direct/indirect (hidden) factors, which contribute to the network evolving, should be traced and identified;
- the trend/principle of the evolving should be identified with appropriate
  approaches and try to make the future status of the evolving networks be
  predictable. This mainly depends on specific application scenarios (datasets);
- local (micro) and global (macro) changes on evolving networks should be
  distinguished and synthesized, so that the dynamics of the evolving networks
  can be examined systematically;
- control and optimization on dynamic networks is important issues to adjust
  the network performance;
- the computing complexity should be controllable and feasible, particularly
  in dealing with large scale and probabilistic data.

Thus more and efficient approaches and strategies should be developed to re-
solve the challenging problems and to improve the dynamic modelling of modern
evolving networks.

References

1. Albert, R., Barabási, A.-L.: Statistical mechanics of complex networks. Rev. Mod.
   Phys. 74(1) (January 2002)
2. Aiello, W., Chung, F., Lu, L.: A random graph model for power law graphs. Exp.
   Math. 10, 53–66 (2001)
3. Barabási, A.-L., Albert, R.: Emergence of scaling in random networks. Science 286(5439), 509–512 (1999)
4. Berzuini, C.: Representing Time in Causal Probabilistic Networks. In: Proceedings
   of the Fifth Annual Conference on Uncertainty in Artificial Intelligence, vol. 10,
   pp. 15–28 (1989)

5. Bilgin, C.C.: Dynamic network evolution: models, clustering, anomaly detection. Technical Report, Rensselaer University, NY (2008)
6. Björkman, K., Karanta, I.: A Dynamic Flowgraph Methodology Approach Based on Binary Decision Diagrams. In: Proceedings of International Topical Meeting on Probabilistic Safety Assessment and Analysis 2011, pp. 267–278 (2011)
7. Bengio, Y., Simard, P., Frasconi, P.: Learning long-term dependencies with gradient descent is difficult. IEEE Transactions on Neural Networks 5, 157–166 (1994)
8. Casteigts, A., Flocchini, P., Quattrociocchi, W., Santoro, N.: Time-varying graphs and dynamic networks. CoRR, abs/1012.0009 (2010)
9. Ding, J.: Probabilistic Fault Management in Distributed Systems. VDI Verlag, Germany (2008) ISSN: 0178-9627, ISBN: 978-3-18-379110-1
10. Ding, J.: Advances in Network Management. CRC Press, Taylor & Francis Group (2010) ISBN-13: 978-1420064520
11. Ding, J., Bouvry, P., Balasingham, I.: Management Challenges for Emerging Wireless Networks. In: Makaya, C., Pierre, S. (eds.) Emerging Wireless Networks: Concepts, Techniques and Applications, pp. 3–34. CRC Press (2012) ISBN: 9781439821350
12. Erdős, P., Rényi, A.: On the evolution of random graphs. Publ. Math. Inst. Hung. Acad. Sci. 5, 17–61 (1960)
13. Gong, S., Xiang, T.: Recognition of group activities using dynamic probabilistic networks. In: ICCV, pp. 742–749 (2003)
14. Harary, F., Gupta, G.: Dynamic graph models. Mathl. Comput. Modelling, 79–87 (1997)
15. Flesch, I., Lucas, P.: Independence decomposition in dynamic Bayesian networks. In: Mellouli, K. (ed.) ECSQARU 2007. LNCS (LNAI), vol. 4724, pp. 560–571. Springer, Heidelberg (2007)
16. Lua, E.K., Crowcroft, J., Pias, M., Sharma, R., Lim, S.: A survey and comparison of peer-to-peer overlay network schemes. IEEE Communications Surveys & Tutorials 7(2), 72–93 (2005)
17. Lin, C.-Y.: Information flow prediction by modeling dynamic probabilistic social network. In: Proc. Int. Conf. Netw. Sci. (2007)
18. Lekuona, A., Lacruz, B., Lasala, P., Fisher, D., Lenz, H.J.: Modeling and monitoring dynamic systems by chain graphs. In: Learning from Data: Artificial Intelligence and Statistics V. Lecture Notes in Statistics, vol. 112, pp. 69–77. Springer-Verlag (1996)
19. Mirowski, P., LeCun, Y.: Dynamic Factor Graphs for Time Series Modeling. In: Buntine, W., Grobelnik, M., Mladenić, D., Shawe-Taylor, J. (eds.) ECML PKDD 2009, Part II. LNCS, vol. 5782, pp. 128–143. Springer, Heidelberg (2009)
20. Mihajlovic, V., Petkovic, M.: Dynamic Bayesian Networks: A State of the Art. Technical Report TR-CTIT-01-34, Centre for Telematics and Information Technology University of Twente, Enschede (2001) ISSN 1381-3625
21. Murphy, K.P.: Dynamic Bayesian Networks: Representation, Inference and Learning. PhD thesis, University of California (2002)
22. Newman, M.E.J.: The Structure and Function of Complex Networks. SIAM Rev. 45, 167–256 (2003)
23. Peköz, E., Röllin, A., Ross, N.: Total variation error bounds for geometric approximation. Bernoulli 19(2), 610–632 (2013)
24. Strogatz, S.: Exploring complex networks. Nature 410, 268–276 (2001)
25. Tucker, A., Liu, X.: Learning dynamic bayesian networks from multivariate time series with changing dependencies. In: Berthold, M., Lenz, H.-J., Bradley, E., Kruse, R., Borgelt, C. (eds.) IDA 2003. LNCS, vol. 2810, pp. 100–110. Springer, Heidelberg (2003)

Part II

Evolutionary Computation for Vision, Graphics, and Robotics

Evolving an Artificial Visual Cortex for Object Recognition with Brain Programming

Gustavo Olague[1,*], Eddie Clemente[1,2], León Dozal[1], and Daniel E. Hernández[1]

[1] Proyecto EvoVisión,
Departamento de Ciencias de la Computación, División de Física Aplicada,
Centro de Investigación Científica y de Educación Superior de Ensenada,
Carretera Ensenada-Tijuana No. 3918, Zona Playitas, Ensenada, 22860, B.C., México
{olague,ldozal,dahernan}@cicese.mx
[2] Tecnológico de Estudios Superiores de Ecatepec, Avenida Tecnológico S/N, Esq. Av. Carlos
Hank González, Valle de Anáhuac, Ecatepec de Morelos
eddie.clemente@gmail.com

Abstract. This chapter describes a new approach to synthesize an artificial visual cortex based on what we call brain programming. Primate brains have several distinctive features that help in the outstanding display of perception achieved by the visual system, including binocular vision, memory, learning, and recognition, to mention only a few. These features are obtained by a complex arrangement of highly interconnected and numerous cortical visual areas. This chapter describes a system composed of an artificial dorsal pathway, or where stream, and an artificial ventral pathway, or what stream, that are fused to create a kind of artificial visual cortex. The idea is to show that brain programming is able to evolve a high number of heterogeneous trees thanks to the hierarchical structure of our virtual brain. Thus, the proposal uses two key ideas: 1) the recognition of objects can be achieved by a hierarchical structure using the concept of function composition, 2) the evolved functions can be discovered through the application of multiple runs of genetic programming that works concurrently using the hierarchical structure. Experimental results provide evidence that high recognition rates could be achieved for a well-known multiclass object recognition problem.

Keywords: Artificial Visual Cortex, Brain Programming, Object Recognition.

1 Introduction

The most sophisticated natural vision system that we know of, is the human visual system; it is capable of recognizing an object even under different variations in location, size, rotation, viewpoint, and lighting conditions. Furthermore, the natural visual system is able to differentiate between visually similar objects and to identify them. Moreover, it is also able to categorize a set of objects by finding common characteristics among them. Hence, trying to mimic these capabilities is a challenging problem that is attracting the evolutionary computer vision community. Thus, in this chapter we describe a new biologically inspired algorithm to approach the object classification

---

* Corresponding author.

O. Schütze et al. (eds.), *EVOLVE - A Bridge between Probability, Set Oriented Numerics,*
and Evolutionary Computation III, Studies in Computational Intelligence 500,
DOI: 10.1007/978-3-319-01460-9_5, © Springer International Publishing Switzerland 2014

problem. Aside from the aforementioned capabilities, the human brain is a great example of a purposeful system, since it transforms a set of complex input signals into a set of complex actions or decisions. Nowadays, the exact manner in which these transformations are performed remains a mystery, and this work attempts to emulate such functionality by evolving an artificial cortex.

According to the neurologist the brain is divided into four lobes: frontal, temporal, parietal, and occipital. The last one is of special interest to the scientific community working in artificial vision, because the visual cortex is located in this lobe. In fact, the primary visual cortex and the secondary visual area are specialized in image processing, for object localization, as well as direction, velocity and trajectory estimation. The advent of computer technology brought new perspectives, where the brain is modeled as an information processing system, by establishing the research field known as computational neuroscience. Although the complexity of the brain is recognized within the domain of evolutionary computation; there is no meaningful work on the development of algorithms modeling the ventral stream and its application to the solution of complex problems [9]. Figure 1 illustrates the main idea that we are proposing to approach the problem of object recognition. We divide the approach into two key ideas. The first one is related to the identification of salient features, through the application of a set of functions that should be able to identify the salience properties that characterize a given object. In general, systems that model the human visual system like [[8],[27], [20],[17],[13]] are data-driven. They are based on a set of patches that are used as a dictionary of visual words. This set of small images represent common and useful characteristics present in all images of a database, integrated by a number of visual categories. The hypothesis in our work is that such set of patches can be replaced by a set of mathematical functions. The second idea for our work is based on the concept of an organ; in biology, an organ is described as a collection of tissues joined in a structural unit to serve a common function. In particular, we are interested in the visual cortex and the functionality of the different tissues involved in the object recognition problem. Hence, the idea is to emulate the functionality of the organ as a set of mathematical functions within a predefined structure and the approach is being called brain programming.

This chapter is organized as follows. This Section describes the biological inspiration and motivation to the present proposal; Section 2 defines how is performed the Artificial Visual Cortex process; in this way, the proposed GP-based methodology called Brain Programming is detailed in Section 3. Afterward, experimental results are presented in Section 4. Finally, the last section contains concluding remarks.

1.1 Statement of the Problem

Object recognition is understood as the problem of determining if an object of a given set is present in a given image or image sequence. Thus, object recognition is considered as an open problem since there is not close solution.

The goal of this chapter is to outline a methodology based on two novel approaches. First, we propose a new model that we called artificial visual cortex. This is implemented through the modeling of a hierarchical structure of the visual cortex together with the concept of function composition, which are inspired from neurological,

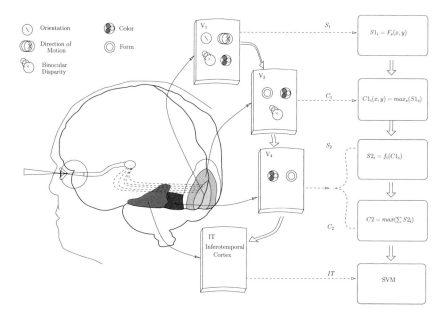

Fig. 1. Analogy between the ventral stream and the proposed computational model

psychological, and physiological disciplines. Second, the artificial visual cortex is synthesized through a new bioinspired strategy that we called brain programming, which mimics the manner of how the visual cortex performs the extraction and description of the visual information needed for the characterization of an object. In other words, the brain programming strategy searches for a set of mathematical expressions that emulate the functionality of specialized tissues present in the brain. In this way, a functional approach is enforced in order to solve the problem of object recognition.

1.2   Visual Cortex

In nature, the visual information processing is done in the brain, and its functionality is described through the concept of two visual subsystems. Nowadays, one of the most accepted descriptions of the visual perception phenomenon is the two-streams hypothesis, which is based on the neuropsychological, neurophysiological, and psychophysical evidence that exhibit the existence of two visual processing regions, known as the ventral and dorsal streams. In this way, the two-stream hypothesis states that both subsystems receive the same visual information as input, but their difference lies on the transformations the information suffers on the streams. This is clearly exposed in the change of paradigm from a what/where dichotomy into a vision-for-action/vision-for-perception duality used to explain the same dorsal/ventral anatomical distinction, see [[25], [26], [32], [31], [19]]. Next, we will give a brief explanation of these two streams.

First, the dorsal stream is mainly related to the spacial location task in the visual processing system, for this reason it is known as the "where" or "how" pathway.

Although, there is still some controversy on the functionality of this pathway, since it is also related to the guidance of actions, as well as, the spacial localization of objects in the scene. The dorsal stream starts at the retina, and it perceives its main input from the magnocellular retinocortical layer of the lateral geniculate nucleus of the thalamus, but it also receives direct subcortical inputs from the superior colliculus and pulvinar structures. First, the visual information is projected on to the $V1$ layer, which is part of the primary visual cortex, also called striate cortex, and it is located on the back side of the brain. Then, the stream continues through the $V2$ and $V3$ layers, followed by the middle temporal area MT, and the medial superior temporal area MST, which are part of the extrastriate visual cortex; and finally the stream lead to the posterior parietal cortex and adjacent areas. In general, it is accepted that the visual attention process is achieved by the dorsal stream, and the most popular paradigm for this process is the feature integration theory presented in [29]. Nevertheless, there are other theories that seek to describe how visual attention is done within the dorsal stream, like [23] and [34]; also, there are some who relate the visual attention process to both streams, see [6]. It is important to note that the first computational model for visual attention was presented by Koch and Ullman in 1985, see [14]. Later, other researchers proposed several methodologies mostly based on the feature integration theory, see [18], and [13]. In all these models the image is decomposed in several dimensions in order to obtain a set of conspicuity maps, which are then integrated into a single map called, the saliency map.

Then, the ventral stream, also known as the "what" pathway, since it is largely associated with object recognition and shape representation [22]. In the same way as the dorsal stream, the ventral stream starts at the retina, and it gets most of its input from the parvocellular layer of the lateral geniculate nucleus of the thalamus, and it is projected onto the $V1$ layer. Then, this stream continues through the $V2$ and $V4$ visual regions that are part of the extrastriate visual cortex. Finally the ventral stream ends at the TEO and TE areas of the inferior temporal cortex. From a computational perspective, the ventral stream is considered a hierarchical, feedforward, and biologically inspired information processing system, specialized in object recognition [12]. The majority of the proposed models for this pathway start by decomposing the image into a set of alternating "S" and "C" layers, named after the simple and complex cells, discovered by Hubel and Wiesel, see [11]. This concept was first implemented by Fukushima in his neocognitron system [8]. This system was later improved by other authors like LeCun with his convolutional networks, and by the HMAX model presented in [24]. In these models, the "S" or simple layers are defined by a set of local filters applied to find higher-order features, and the "C" complex layers increase the features invariance by combining units of the same kind, see [30].

In summary, the visual system has been defined by two information processing streams organized in two broad structures subserving object recognition and spatial vision. The classical dichotomy between object and space perception focuses on the importance of a single and general purpose representation. On the other hand, the "what" and "how" theory of Milner and Goodale [19] gives emphasis to the idea that the visual system is defined according to the requirements of the task that each stream subserves. The idea is to define multiple frames of reference giving special attention to the goal of the observer. In this way, the same object and spatial information is transformed by the visual system for different purposes. Thus, the ventral system represents the visual world in allocentric coordinates by promoting conscious perceptual awareness. On the other hand, the dorsal stream uses egocentric coordinates to transform the information about objects location, orientation, and size; see [5].

1.3 Evolution and Teleology for Visual Processing

This section describes the idea that visual processing is the result of the evolution of the brain; hence, the application of artificial evolution for generating object recognition algorithms is coherent with that idea. The computational approach for our work can be described in two phases. First, we studied some theory about the evolution of the brain. Then, we propose a teleological explanation for the two stream hypothesis in the natural visual system. Nowadays, there are two main schools of knowledge, teleological and mechanistic, they both seek to explain how natural systems work. Nevertheless, it is important to note that teleological explanations do not deny the mechanistic aspects of a system. Although, there is a controversy due to the fact that teleological explanations are not able to materialize the concept of purpose, and on the other hand, mechanistic explanations cannot avoid the idea of purpose when explaining a system. Moreover, a purpose is not a desire since when we talk about a purpose we need to ask if it was achieved or not. Therefore, we propose that a teleological viewpoint expands the concept of artificial evolution by bringing new and richer explanations of how evolution works. This viewpoint corresponds to many ideas about the evolution of the human brain which are described in a richer way through a teleological perspective, see [19,5,2,1,16,28,21].

The classical conception of the visual system assumes that it constructs some kind of internal model of the environment. In other words, that the system builds a representation of the real world inside the brain, which serves as the frame of reference for all the visual driven actions and decisions. The idea of studying the structure and functionality of the neocortex is based on several explanations: comparative, developmental, and functional or adaptationist. In particular, we share the doctrine of several biologists, see [16,28,2], that adaptation in nature, specializes the organs to cover certain necessities of the organism; hence, developmental and functional explanations are complementary and not alternate explanations. This is confirmed by the fact that some species from the same taxonomic group have evolved specialized visual mechanisms, which may work in distinct manners; but are coherent and correlated to a given particular function of the organism. Moreover, most scientists agree that such structures are the result of the evolution using the fundamental principle of natural selection [2,1,16,28,21]. Thus, the objective of this chapter is to evolve a specialized system based on the functionality of the ventral and dorsal streams, adapting its behavior to an specific task, in particular for solving the object classification problem.

2 Artificial Visual Cortex (AVC)

The Artificial Visual Cortex model is based on the idea of an organ, which is defined as a collection of tissues joined in a structural unit to serve a common function. Thus, the

AVC is a model of the visual cortex and the way the visual information is processed. The different models reported on the state-of-the art are inspired from the natural visual system focusing the description of the dorsal and ventral streams as separate subsystems; the specific goals aim to solve the visual attention or object recognition tasks, respectively. It is to note the lack of a general work that attempts to model the visual cortex as a whole and unique system. In general, there are only few works studying these aspects where the visual attention output is considered as an input to the object recognition module in order to create a more complex system. Instead of this simple approach, and in concordance with the ideas reported in [19,5] we would like to emphasize that the layers $V1$ and $V2$ are both part of the dorsal and ventral streams. Hence, we propose here a new approach that results in an artificial visual cortex model using the idea of a common functionality produced at the early stages of the dorsal and ventral pathways, see Figure 2. In this way, the AVC offers the functionality of the dorsal stream of distinguishing early regions in the image, which are used to describe an object and later continuing it with a model of the ventral stream. Note, that for this new methodology it is necessary to understand an image as the first input to the visual system. Due to the nature of function composition between layers in the brain, we propose to define an image as the graph of a function, see [21]. The function in this case is the base for understanding the transformation of the physical, geometrical, or other properties of the scene.

Definition 1 (Image as the graph of a function). *Let f be a function $f : U \subset \mathbb{R}^2 \to \mathbb{R}$. The graph or image I of f is the subset of $\mathbb{R}^3$ that consist of the points $(x, y, f(x,y))$, in which the ordered pair (x,y) is a point in U and $f(x,y)$ is the value at that point. Symbolically, the image $I = \{(x, y, f(x,y)) \in \mathbb{R}^3 | (x,y) \in U\}$.*

In this way, the image, seen as the graph of a function, is the input of a computational system that mimics the functionality and hierarchical structure of the natural visual system. In other words, each layer of the visual cortex can be modeled through a set of mathematical functions that represent a virtual tissue. In this work, the object recognition system is designed following a function driven paradigm. Contrary to previous research [[8],[27], [20],[17],[13]] where the ventral stream is modelled through a data-driven scenario.

   Thus, the paradigm of genetic programming is used to implement the proposed approach for which a set of evolutionary visual operators (EVOs) are optimized according the functionality of each tissue following the hierarchical structure of the AVC. The final result is the optimal object recognition program that satisfies the object recognition task. Hence, the aim of genetic programming is to find the best set of EVOs using a number of functions over the image working as building blocks over the whole hierarchical structure. The goal in this work is to find a solution to a multiclass object recognition problem. A result of the functional approach compared to previous, data-driven approaches is reflected on the lower amount of computations that brings a significant economy in the number of operations without sacrificing the overall quality. Hence, the EVO is defined as follows:

Definition 2. *EVO*. *The Evolutionay Visual Operators (EVO) should be understood as a general concept that is applied to the evolution of brains; in such a way, that each*

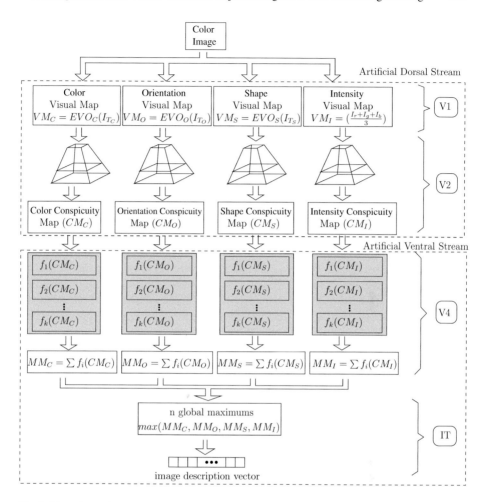

Fig. 2. Flowchart of the artificial visual cortex. Note, the similarity with the visual attention process in which the image is decomposed into several dimensions. In our approach a function driven paradigm is enforced to avoid the application of image patches.

step of the visual information processing is replicated by specialized programs that fit the functionality of an artificial tissue in an optimal way.

Next, the artificial dorsal and ventral streams are described.

2.1 Artificial Dorsal Stream (ADS)

The ADS is based on the feature-integration theory for attention, proposed by Treisman and Gelade in 1980, see [29], that suggests that attention must be processed serially by each stimulus; since, the display of more than one separable feature should be coordinated to characterize or distinguish the possible objects in the scene. In this way, the first step of the ADS is represented by the image acquisition stage. Here, the system considers the digital color images, in the RGB color space, and the transformation into the CMYK, and HSV color spaces, as the input to the algorithm. Thus, the color image is defined as a set of components from each color space $I_{color}=\{R,G,B,C,M,Y,K,H,S,V\}$. Next, three evolved visual operators are applied separately to emphasize features based on color, orientation, and shape. In biological models, that use a data-driven approach, operators such as opponency color [13], or Gabor filters [27] have been proposed according to the knowledge from neuroscience based on how these features are obtained in the visual cortex of the brain. Here, the operations of the ADS are evolved with genetic programming to obtain an optimal set of EVOs as depicted on Figure 2. The EVO is a specialized function that was evolved from a set of image operators; and each EVO has suitable characteristics that are used to create a set of visual maps along the feature dimension of color, orientation and shape. Then, for each visual map an image pyramid is created in order to achieve position and scale invariance. Thus, the pyramid is reduced to a conspicuous map for each of the aforementioned feature dimensions. In this way, it is said that artificial evolution is in charge of optimizing the three functions that extract the information in color, orientation, and shape dimensions, resulting in an EVO for color EVO_C, orientation EVO_O, and shape EVO_S. These aspects of the optimization approach enforce the solution of achieving invariance through the definition of the objective function. Moreover, the hierarchical structure helps to achieve the desired result using the concept of function composition. Next, we describe the color, orientation and shape visual maps.

Color Visual Map. The idea of building a color visual map VM_C, see Figure 2, is to create an image that contains outstanding information along the color feature dimension of the image. Thus, the input image is transformed by the function, $EVO_C : I_{color} \rightarrow VM_C$, to enhance the color features. In this way, an EVO_C is evolved with tree-based genetic programming in order to extract the most significance characteristics of the image through the color feature dimension. Hence, the evolutionary process uses the set of functions and terminals shown in Table 1 to generate the operators for the color dimension. The notation is summarized as follows: A, B can be any of the terminals, as well as the output of any of the functions, or a composition of them; I is the image that enters the AVC.

Orientation Visual Map. The orientation visual map VM_O is produced by applying $EVO_O : I_{color} \rightarrow VM_O$, this operator is evolved with tree-based genetic programming to optimize the extraction of edge and corner information within the input image I or in its color components I_{color}. In this way, the elements values of the visual map VM_O represent the feature prominence of the orientation dimension. Thus, the brain programming applies the functions and terminals of Table 2, in order to enhance the best orientation features that are useful for the object recognition task. The notation used is as follows: A or B can be any of the terminals in the Table 2; as well as, the output of any of the functions; D_u symbolizes the image derivatives along direction $u \in \{x,y,xx,yy,xy\}$; G_σ are Gaussian smoothing filters with $\sigma = 1$ or 2. Note that, the terminal I represents the image in gray scale and that it could also be the representation of the color component V in the HSV color space.

Table 1. Set of functions and terminals used by EVO_C to create the visual map VM_C

Color Functions	Description
$A+B, A-B, A \times B, A/B$	Arithmetic functions between two images A and B
$log(A), exp(A)$	Transcendental functions over the image A
$(A)^2$	Square function over the image A
$\sqrt{A}$	Square root function over the image A
A^c	Image complement
$round(A), \lfloor A \rfloor, \lceil A \rceil$	Round, floor and ceil functions over the image A
$thr(A)$	Dynamic threshold function over the image A

Color Terminals	Description
R,G,B	Color components of the RGB color model
C,M,Y,K	Color components of the CMYK color model
H,S,V	Color components of the HSV color model
$O_{PR-G}(I), O_{PB-Y}(I)$	Color opponency Red - Green and Blue - Yellow

Shape Visual Map. The function used to compute the shape visual map, $EVO_S : I_{color} \rightarrow VM_S$, is evolved with genetic programming to extract the shape information in the input image. In this way, the shape visual map provides characteristics as the form and structure of the interest object within the image. Note, that the proposed functions in Table 3 are obtained by applying mathematical morphology to the image. Thus, according to the literature, the work reported in this chapter could be considered as the first to use morphological image processing for the modeling of the visual cortex.

The intensity visual map is performed by applying the function $VM_I : I_{color} \rightarrow I$; in this case, VM_I is defined as follows: $VM_I = \frac{R+G+B}{3}$. Where R, G and B are the color components of the RGB color model.

Conspicuity Maps. The conspicuity maps (CMs) are obtained by means of a center-surround function, which is applied to the visual maps in order to simulate a set of center-surround receptive fields. The natural structure allows the ganglion cells to measure the differences between firing rates in center (c) and surroundings (s) of ganglion cells. For the artificial vision; first, a pyramid $VM_I(\alpha)$ of nine spatial scales $S = \{1,2,...,9\}$ is created for each of the four resulting VMs. Afterwards, an across-scale substraction $\ominus$ is performed, resulting in a center-surround map $VM_I(\omega)$ in such a way that the value of the pixel is augmented as long as the contrast is increased within their neighbors at different scales. Finally, the $VM_I(\omega)$ maps are added using an across-scale addition $\oplus$ in order to obtain the desired conspicuity maps CM_I.

Until this stage, we have four CMs, one for each feature, as shown in Figure 2. The CMs are obtained similar to Walther and Koch model, see [33]. Next, instead of combining the CMs into a single saliency map, the idea here is to use the four CMs as input to an artificial ventral stream in order to derive a vector descriptor. Thus, the information just obtained will be used by a classification process. In fact, the fitness function proposed in this work is computed from the accuracy achieved with a support vector machine (SVM).

Table 2. Set of functions and terminals used by EVO_O to create the visual map VM_O

Orientation Functions	Description
$A+B, A-B, A\times B, A/B$	Arithmetic functions between two images A and B
$|A|, |A+B|, |A-B|$	Absolute value applied to A, and the adition and substraction operators
$log(A)$	Transcendental functions over the image A
$(A)^2$	Square function over the image A
$\sqrt{A}$	Square root function over the image A
$k+A, k-A, k\times A, A/k$	Arithmetic functions between an images A and a constant k
$round(A), \lfloor A \rfloor, \lceil A \rceil$	Round, floor and ceil functions over the image A
$inf(A,B), sup(A,B)$	Infimum and supremum functions between the images A and B
$G_{\sigma=1}(A), G_{\sigma=2}(A)$	Convolution of the image A and a Gaussian filter with $\sigma=1$ or 2
$D_x(A), D_y(A)$	Derivative of the image A along direction x and y
$thr(A)$	Dynamic threshold function over the image A

Orientation Terminals	Description
$I, D_x(I), D_y(I), D_{yy}(I), D_{xx}(I), D_{xy}(I)$	Gray image scale and its derivatives
$R, D_x(R), D_y(R), D_{yy}(R), D_{xx}(R), D_{xy}(R), G, D_x(G), D_y(G), D_{yy}(G), D_{xx}(G), D_{xy}(G), B, D_x(B), D_y(B), D_{yy}(B), D_{xx}(B), D_{xy}(B)$	Color components of the RGB color model and its derivatives
$C, D_x(C), D_y(C), D_{yy}(C), D_{xx}(C), D_{xy}(C), M, D_x(M), D_y(M), D_{yy}(M), D_{xx}(M), D_{xy}(M), Y, D_x(Y), D_y(Y), D_{yy}(Y), D_{xx}(Y), D_{xy}(Y), K, D_x(K), D_y(K), D_{yy}(K), D_{xx}(K), D_{xy}(K)$	Color components of the CMYK color model and its derivatives
$H, D_x(H), D_y(H), D_{yy}(H), D_{xx}(H), D_{xy}(H), S, D_x(S), D_y(S), D_{yy}(S), D_{xx}(S), D_{xy}(S)$	Color components of the HSV color model and its derivatives

2.2   Artificial Ventral Stream (AVS)

Once, that artificial dorsal stream has highlighted the color, orientation, shape and intensity features in the image; the artificial dorsal stream is charged to describe such important regions. The common approach uses a template matching technique to build a descriptor that characterize the image, this technique implements the correlation between the information obtained with an interest region selection process, and a number of prototype patches. The goal is to identify the successful patches and build an array that is known as the universal dictionary of features [30,20,27], and which is used as an input to SVM classifier. From another point of view, in this chapter the template matching technique is substituted with a functional approach; in other words, the

Table 3. Set of functions and terminals used by EVO_S to create the visual map VM_S

Shape Functions	Description
$A+B, A-B, A \times B, A/B$	Arithmetic functions between two images A and B
$k+A, k-A, k \times A, A/k$	Arithmetic functions between an images A and a constant k
$round(A), \lfloor A \rfloor, \lceil A \rceil$	Round, floor and ceil functions over the image A
$Dil_d(A), Dil_s(A), Dil_{dm}(A)$	Dilation operator with disk, square, and diamond structures
$Erd_d(A), Erd_s(A), Erd_{dm}(A)$	Erosion operator with disk, square, and diamond structures
$Sk(A)$	Skeleton operator over the image A
$Prm(A)$	Find perimeter of objects in the image A
$HM_{M_d}(A), HM_{M_s}(A), HM_{M_{dm}}(A)$	Hit or miss transformation with disk, square, and diamond structures
$T_H(A), B_H(A)$	Performs morphological bottom-hat and top-hat filtering over the image A
$open(A), close(A)$	Open and close morphological operators on A
$thr(A)$	Dynamic threshold function over the image A
Shape Terminals	**Description**
R,G,B	Color components of the RGB color model
C,M,Y,K	Color components of the CMYK color model
H,S,V	Color components of the HSV color model

artificial ventral stream uses a set of evolved functions that are capable of replacing the universal dictionary, and we claim that it corresponds to a function driven approach. These functions enhancing the set of prominent features that were emphasized during the interest region detection computed in previous stages. Thus, the computational operations in comparison with a template matching technique are minimized due to the definition of the descriptor as set of functions. According to Figure 2, the information provided by the conspicuity maps is feedforward to k operators that emulate a set of lower order hypercomplex cells replicating the functionality of a virtual tissue. Hence, all evolved functions along each dimension, color, orientation, shape and intensity, are added in order to obtain an image that we called mental map MM_i, over each feature. In this way, all mental maps are combined with a max operation, that is used to select the most n higher values and build an array which is the image descriptor. Note, that each function is an evolved visual operator (EVO) built by the organic genetic programming from the particular set of terminals and functions shown in Table 4. Note also that this

Table 4. Set of functions and terminals for the ventral stream to create the mental maps MM_i

Functions	Description
$A+B, A-B, A \times B, A/B$	Arithmetic functions between two images A and B
$\|A+B\|, \|A-B\|$	Absolute value applied to the addition and substraction operators
$log(A)$	Transcendental functions over the image A
$(A)^2$	Square function over the image A
$\sqrt{A}$	Square root function over the image A
$G_{\sigma=1}(A), G_{\sigma=2}(A)$	Convolution of the image A and a Gaussian filter with $\sigma = 1$ or 2
$D_x(A), D_y(A)$	Derivative of the image A along direction x and y
Terminals	Description
$CM, D_x(CM), D_y(CM), D_{yy}(CM), D_{xx}(CM), D_{xy}(CM)$	Conspicuity Maps and its derivatives

second stage could be said to perform an information description operation (IDO) with the aim of discovering the best set of functions that creates the most discriminant vector of characteristics. Finally, the descriptor is the input of a SVM classifier that after a training process decide over the image class membership.

3   Synthesizing AVCs through Brain Programming

This section describes the characteristics of the evolutionary process used to synthesize AVCs, through the application of brain programming. Thus, the artificial brain is composed of different artificial tissues or layers, which perform a set of functions towards a single objective. Of course, the proposed methodology could be easily enhanced for multiobjective tasks. In this sense, the AVC is represented by a complex chromosome, defined as an array of functions where each of them corresponds to a virtual tissue. These functions are written as a tree structure and can be interpreted as the genes from an evolutionary point of view. The phenotype is encoded as depicted in Figures 2, and 3, such diagrams illustrate the complexity of the proposed system using a kind of heterogeneous and hierarchical genetic programming. Thus, the AVC should be seen as a single entity, with the objective of classifying several objects. In other words, the functional representation of an artificial organ is designed as a hierarchical and heterogeneous structure, according to the task at hand. In this way, the size of the chromosome can change as well as the depth of the genes, throughout the whole evolutionary process. Hence, it is important to note that each individual in the population should be understood as a complete system and not only as a list of tree-based programs. This complex information processing system executes the following steps.

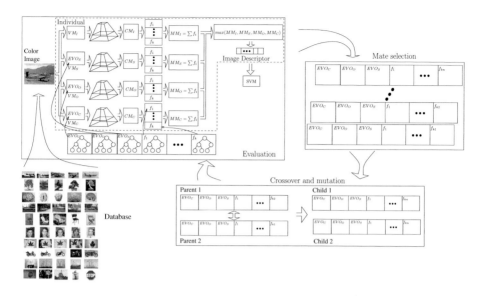

Fig. 3. General flowchart of the methodology to synthesize an artificial visual cortex

1. Randomly generate an initial population with a ramped half-and-half technique and a maximum depth of 7 levels for each gene. The size of the whole chromosome is randomly initialized with a maximum length of 15 genes.
2. The evolutionary loop starts the execution of each AVC by computing its fitness using a SVM that is used to calculate the classification rate for a given training image database.
3. A set of AVCs is selected from the population with a probability based on fitness using a roulette-wheel selection to participate in the genetic recombination; and the best AVC is retained for further processing.
4. A new individual of the population is created from the selected AVC by applying the genetic operators using the crossover or mutation operation at chromosome or gene levels. Similar to genetic algorithms, brain programming executes the crossover between two selected AVCs at the chromosome level by applying a "cut-and-splice" crossover, see Figure 4. Thus, all data beyond the selected crossover point in either AVC string is swapped between both parents A and B. On the other hand, the result of applying a crossover at the gene level is performed by randomly selecting two AVC parents based on fitness in order to execute a subtree crossover between both selected genes. Moreover, the mutation at the chromosome level leads the selection of a random gene of a given parent to replace such substructure by a new randomly mutated gene. Thus, the mutation at the gene level is calculated over an AVC by applying a subtree mutation to a probabilistically selected gene; in other words, a mutation point is probabilistically chosen at a selected gene and the subtree up to that point is removed and replaced with a new subtree.

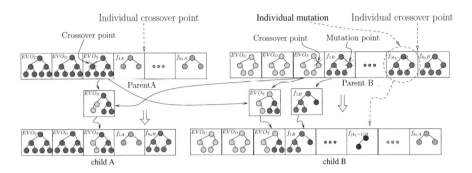

Fig. 4. The genetic operations are applied at the gene and chromosome levels

Table 5. This table shows a comparison in the performance between HMAX and AVC. Note, that in the case of the HMAX model a learning process was necessary to identify the best patches, while for the AVC the brain programming was used to discover the best *EVOs*.

Data sets	Performance of HMAX		Artificial
	boost	SVM	V. C.
Airplanes	96.7 %	94.9 %	100 %
Cars	95.1 %	93.3 %	99 %
Faces	98.2 %	98.1 %	100 %
Leaves	97.0 %	95.9 %	100 %
Motorbikes	98.0 %	97.4 %	100 %

5. The evolutionary loop is terminated until an acceptable classification rate is reached; i.e., the accuracy is equal to 100%, or the total number of generations $N = 30$ is reached.

4   Experimental Results

This section provides details about the experiments in order to explain the system that was implemented to learn an artificial visual cortex. All experiments were performed in a Dell Precision T7500 Workstation, Intel Xeon 8 Core, NVIDIA Quadro FX 3800 and Linux OpenSUSE 11.1 operating system. In this way, the SVM Matlab implementation developed by Chan and Lin, see [3], was used in order to compare with the HMAX model [27]. Thus, two different experiments were designed, the first one develops a binary test, where the performance of the proposed model in the object present/abscent experiment using five object classes was evaluated; for this experiment the classes: airplanes, cars-rear, faces, leaves, motorbikes, and the background used as negative class from the Caltech 5, see [7]. Table 5 shows a comparison with the HMAX model using boost and SVM as classifiers; this last one was used in the test performance of the AVC after the evolution process.

Table 6. Individuals found after the evolutionary process

Name	EVO	IDO	Evaluation
Airplanes Vs Bg	$EVO_O = log(D_{yy}(H))$ $EVO_C = \sqrt{(Y^2 - M/C)^2}$ $EVO_S = Erd_s(H - S)$	$f_1 = 0.25 * (I)$ $f_2 = log(D_x(I))$	$Tr = 100\%$ $Val = 100\%$ $Tst = 100\%$
Cars Vs Bg	$EVO_O = 0.5(Y)$ $EVO_C = Op_{B-Y}(I)$ $EVO_S = M$	$f_1 = G_{\sigma=2}(D_{yy}(I))$ $f_2 = D_{xx}(I)$ $f_3 = G_{\sigma=2}(G_{\sigma=1}(D_y(I)))$ $f_4 = G_{\sigma=1}(\|Dy(I)\|)$	$Tr = 100\%$ $Val = 100\%$ $Tst = 98\%$
Faces Vs Bg	$EVO_O = G_{\sigma=2}(D_{xx}(I))$ $EVO_C = log(Op_{R-G}(I))$ $EVO_S = thr(R)$	$f_1 = \sqrt{D_{xx}(I)}$	$Tr = 99.29\%$ $Val = 100\%$ $Tst = 100\%$
Leaves Vs Bg	$EVO_O = G_{\sigma=2}(\|D_x(G)\|)$ $EVO_C = \sqrt{(Op_{B-Y}(I))^c}$ $EVO_S = R - G$	$f_1 = 0.5 * (D_{xx}(I))$ $f_2 = log(D_{xxx}(I))$	$Tr = 97.5\%$ $Val = 100\%$ $Tst = 100\%$
Motorbikes Vs Bg	$EVO_O = \|\lceil D_{xx}(C)\rceil - inf(D_x(G),H)\|$ $EVO_C = log((Op_{R-G}(I))^c)$ $EVO_S = H_{M_d}(H) - 0.56$	$f_1 = D_x(G_{\sigma=2}(D_y(I)))$ $f_2 = D_y(G_{\sigma=2}(D_{xx}(I)))$ $f_3 = \sqrt{D_{xx}(I)} + D_x(I)$ $f_4 = D_x(G_{\sigma=2}(D_{xx}(I)))$	$Tr = 100\%$ $Val = 100\%$ $Tst = 100\%$

In this experiment, each data set was randomly divided into three sets, training, validation and testing. The number of positive training image was variable (10, 20, 30, 40, 50, 60 ,70) and the negative image was 50. For validation and testing was used 50 negative and positive images. In all cases, the function fitness was defined by the accuracy in the validation set. The performance is depicted in Figure 5.

Note that, the 100% accuracy in testing was achieved by AVC with an exception for the class cars-rear with an accuracy of 99%; nevertheless, this perfomance is higher than the one achieved by the HMAX. In the other cases, the 100% was scored with 10 and 40 images of training for the airplanes; in the case of faces, the 100% in testing was achieved with 70 images in training; for the class leaves the 100% perfomance was achieved with 20 images in training; and for the motorbikes with 40 images in training was reached the 100% in testing. These individuals are shown in Table 6, and Figure 6 depicts the fitness behavior along the evolution; while Figure 7 presents the complexity and diversity for the best individual for each generation. Thus, Figures 7 (a) and (b) describe the complexity of the best individual indicated in Table 6. The Figure 7(c) shows the amount of genetic diversity found in the population at each generation along the run that produces the fittest individual in the airplanes vs background experiment. In this way, the complexity of the best individual is quantified by the three deepth, see Figure 7 (a), and the number of nodes, see Figure 7 (b); and the diversity is defined as the percentage of operators uniqueness within the population. Note that, the best individual is a structure of EVOs and the complexity and diversity is over each operator along the evolution. Hence, the Figures 7 (d), (e), and (f) describe the complexity and diversity over the experiment cars vs background. The Figures 7 (g), (h), and (i) correspond to the complexity and diversity of the evolution of the best individual in the faces vs background case. The Figures 7 (j), (k), and (l) describe the complexity and diversity of the evolution of the fittest individual in the leaves vs background experiment. Finally; Figures 7 (m), (n), and (o) depict the complexity and diversity in the motorbikes vs background experiment.

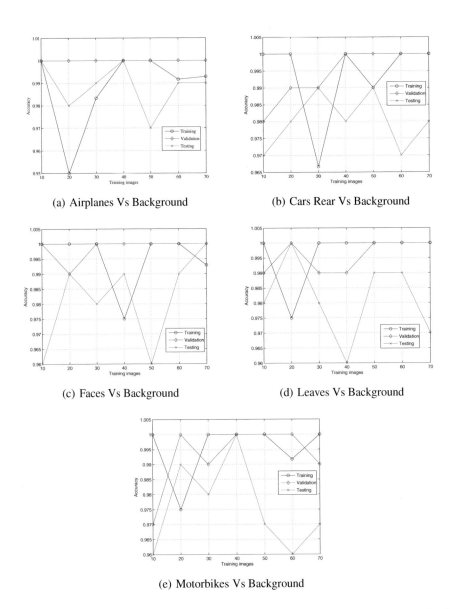

(a) Airplanes Vs Background

(b) Cars Rear Vs Background

(c) Faces Vs Background

(d) Leaves Vs Background

(e) Motorbikes Vs Background

Fig. 5. Performance achieved by the AVC varying the training set for (a) Airplanes Vs Background, (b) Cars Side Vs Background, (c) Faces Vs Background, (d) Leaves Vs Background, and (e) Motorbikes Vs Background

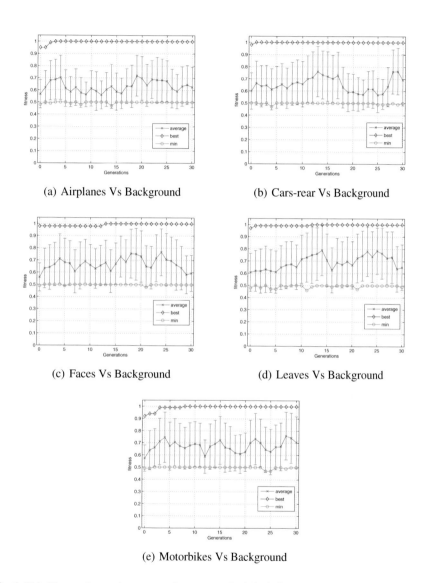

(a) Airplanes Vs Background

(b) Cars-rear Vs Background

(c) Faces Vs Background

(d) Leaves Vs Background

(e) Motorbikes Vs Background

Fig. 6. This Figure shows the average fitness, standard deviation, best, and minimum fitness of the run that produces the best individual for the airplanes, cars-rear, faces, leaves, and motorbikes classes; discussed in Table 6

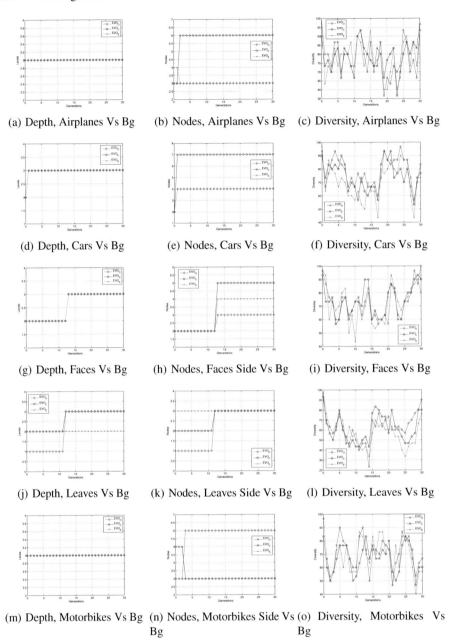

(a) Depth, Airplanes Vs Bg   (b) Nodes, Airplanes Vs Bg   (c) Diversity, Airplanes Vs Bg

(d) Depth, Cars Vs Bg   (e) Nodes, Cars Vs Bg   (f) Diversity, Cars Vs Bg

(g) Depth, Faces Vs Bg   (h) Nodes, Faces Side Vs Bg   (i) Diversity, Faces Vs Bg

(j) Depth, Leaves Vs Bg   (k) Nodes, Leaves Side Vs Bg   (l) Diversity, Leaves Vs Bg

(m) Depth, Motorbikes Vs Bg   (n) Nodes, Motorbikes Side Vs   (o) Diversity, Motorbikes Vs
                                   Bg   Bg

Fig. 7. The Figure (a) and (b) describe the complexity of the best individual indicated in the Table 6. The Figure (c) shows the amount of genetic diversity found in the population at each generation along the run that produces the fittest individual in the airplanes vs background experiment for the first row. In this way, the Figure (a) describes the tree depth of the fittest individual along to the evolution and the Figure (b) shows the number of nodes of the best individual. Thus, the complexity and diversity for the cars, faces, leaves and motorbikes classes are also shown in the rows 2, 3, 4, and 5.

Table 7. This table shows a comparison of performances that were obtained with the HMAX, HMAX-CUDA, AVS and AVC

	Image size	HMAX MATLAB	HMAX CUDA	Artificial V. S.	Artificial V. C.
Running time over different image size	896×592	34s	3.5s	2.6s	9.91s
	601×401	24s	2.7s	1.25s	5.32s
	180×113	9s	1s	0.23s	0.49s
Performance over 15 training images per 10 classes		94%	94%	78%	85.3%
Performance over 15 testing images per 10 classes		73%	73%	80%	84%
Number of convolutions		4848	4848	216	95

In the second experiment, the system was tested using 10 classes and 15 images per class of the Caltech 101 database, see [7]. In order to compare with the original HMAX model [27] whose source code was used in our experiments; we provide the Table 7 that presents a summary of the best results and a comparison with the HMAX model; as well as an implementation of the HMAX-CUDA and a previous proposal called the artificial ventral stream (AVS), see [4]. Note, that the total number of convolutions is much lower than the HMAX and HMAX-CUDA. This aspect is important since the factor of improvement is on the order of a hundred of operations. However, the performance of the AVC is lower than the HMAX model, but its level is worse in testing, while the effectiveness of our approach remains constant. Figure 9 shows the average fitness of the run where the best program was obtained, while Figure 8 depicts the process of this individual applied to an image. Also, Figure 10 illustrates the range of descriptor values of the best solution for each class. We provide also the overall results of the best AVC through the confusion matrix, see Table 8.

Table 8. This table shows the results of the best solution in the form of a confusion matrix obtained during the testing of the AVC. The final accuracy $acc = 84\%$ classifies correctly (126/150) images.

	Airplanes	Bonsai	Brains	Cars	Chairs	Faces	Leaves	Motorcycle	Schooner	Stop Signal
Airplanes	15	0	0	0	0	0	0	0	0	0
Bonsai	0	9	2	0	3	0	0	0	0	1
Brains	0	2	11	0	1	0	0	0	1	0
Cars	0	0	0	14	0	0	0	0	1	0
Chairs	0	0	2	1	11	0	0	0	0	1
Faces	0	0	0	0	0	14	0	0	1	0
Leaves	0	0	0	0	0	0	15	0	0	0
Motorcycle	0	0	0	0	0	0	0	15	0	0
Schooner	0	1	0	0	0	0	0	0	12	2
Stop Signal	0	2	1	0	0	1	0	0	1	10

116   G. Olague et al.

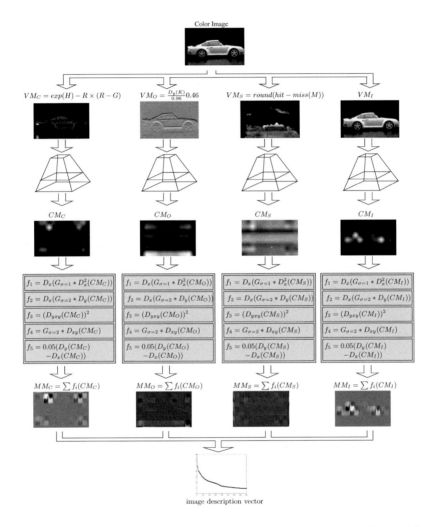

Fig. 8. Flowchart of the best artificial visual cortex discovered with the proposed methodology

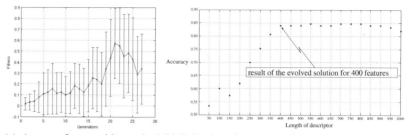

(a) Average fitness with standard deviation

(b) Behavior of accuracy with respect to the descriptor length

Fig. 9. Figure (a) shows the average fitness and standard deviation of the run that produces the best individual. Figure (b) depicts the performance after changing the descriptor length.

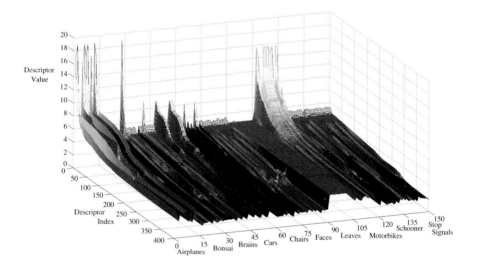

Fig. 10. This plot shows the descriptors that are used as input to the SVM of the best individual

5 Conclusions

In this work the artificial visual cortex was described as a novel methodology inspired from the natural visual system, emulating its hierarchical structure and functionality in order to solve the problem of object classification. In this manner, a new evolutionary approach, called brain programming, was designed to synthesize an artificial visual cortex specialized for the object recognition task. This approach allows the evolution of a group of independent visual operators called EVOs. Note that each of them is built with different set of terminals and operators working independently towards a specific goal. Thus, the combination of these operators permits the achievement of a common objective such as the object recognition task. In this way, each evolved visual operator mimics the functionality of a tissue within the artificial brain. Final results achieved by the system score a classification accuracy of 99% for the two-class test while accuracy rates in a multi-class test were similar to the state-of-the-art. Nevertheless, the proposed functional approach provides a significant computational simplification of the overall structure leading to a lower computational cost.

Acknowledgements. This research was founded by CONACyT through the Project 155045 - "Evolución de Cerebros Artificiales en Visión por Computadora". Second author supported by scholarship 164884 from CONACyT. This research was also supported by TESE through the project DIMI-MCIM-004/08.

References

1. Ayala, F.J.: Teleological Explanations in Evolutionary Biology. Philosophy of Science 37(1), 1–15 (1970)
2. Barton, R.A.: Visual specialization and brain evolution in primates. Proceedings of the Royal Society of London Series B-Biological Sciences 265(1409), 1933–1937 (1998)
3. Chang, C.C., Lin, C.J.: LIBSVM: A Library for Support Vector Machines. ACM Transactions on Intelligent Systems and Technology 2, 27:1–27:27 (2011), Software available at http://www.csie.ntu.edu.tw/~cjlin/libsvm
4. Clemente, E., Olague, G., Dozal, L., Mancilla, M.: Object Recognition with an Optimized Ventral Stream Model Using Genetic Programming. In: Di Chio, C., et al. (eds.) EvoApplications 2012. LNCS, vol. 7248, pp. 315–325. Springer, Heidelberg (2012)
5. Creem, S.H., Proffitt, D.R.: Defining the cortical visual systems: "what", "where", and "how". Acta Psychologica 107, 43–68 (2001)
6. Desimone, R., Duncan, J.: Neural mechanisms of selective visual attention. Annu. Rev. Neurosci. 18, 193–222 (1995)
7. Fei-Fei, L., Fergus, R., Perona, P.: Learning generative visual models from few training examples: An incremental Bayesian approach tested on 101 object categories. Computer Vision and Image Understanding 106(1), 59–70 (2007)
8. Fukushima, K.: Neocognitron: A self-organizing neural network model for a mechanism of pattern recognition unaffected by shift in position. Biological Cybernetics 36(4), 193–202 (1980)
9. Holland, J.H.: Complex Adaptive Systems. Daedalus 121(1), 17–30 (1992)
10. Hoquet, T.: Darwin teleologist? Design in the orchids. Comptes Rendus Biologies 333(2), 119–128 (2010)
11. Hubel, D.H.: Exploration of the primary visual cortex, 1955-78. Nature, 515–524 (October 1982)
12. Hubel, D.H., Wiesel, T.N.: Receptive fields of single neurones in the cat's striate cortex. J. Physiol. 148(3), 574–591 (1953)
13. Itti, L., Koch, C.: Computational modelling of visual attention. Nature Review Neuroscience 2(3), 194–203 (2001)
14. Koch, C., Ullman, S.: Shifts in selective visual attention: towards the underlying neural circuitry. Hum. Neurobiol. 4(4), 219–227 (1985)
15. LeCun, Y., Bottou, L., Bengio, Y., Haffner, P.: Gradient-based learning applied to document recognition. Proceedings of the IEEE 86(11), 2278–2324 (1998)
16. Lennox, J.G.: Darwin was a teleologist. Biology and Philosophy 8(4), 409–421 (1993)

17. Mel, B.W.: Seemore: Combining Color, Shape, and Texture Histogramming in a Neurally Inspired Approach to Visual Object Recognition. Neural Computation 9(4), 777–804 (1997)
18. Milanese, R.: Detecting salient regions in an image: from biological evidence to computer implementation. PhD thesis, Department of Computer Science, University of Genova, Switzerland (December 1993)
19. Milner, A.D., Goodale, M.A.: The Visual Brain in Action, 2nd edn. Oxford University Press, Oxford (2006)
20. Mutch, J., Lowe, D.G.: Object Class Recognition and Localization Using Sparse Features with Limited Receptive Fields. Int. J. Comput. Vision 80, 45–57 (2008)
21. Olague, G.: Evolutionary Computer Vision – The First Footprints (to appear)
22. Oram, M.W., Perrett, D.I.: Modeling visual recognition from neurobiological constraints. Neural Networks 7(6), 945–972 (1994)
23. Rensink, R.A.: The Dynamic Representation of Scenes. Visual Cognition 7(1-3), 17–42 (2000)
24. Riesenhuber, M., Poggio, T.: Hierarchical models of object recognition in cortex. Nature Neuroscience 2, 1019–1025 (1999)
25. Schneider, G.E.: Contrasting Visuomotor Functions of Tectum and Cortex in the Golden Hamster. Psychologische Forschung 31(1), 52–62 (1967)
26. Schneider, G.E.: Two Visual Systems. Science 163(3870), 895–902 (1969)
27. Serre, T., Kouh, C., Cadieu, M., Knoblich, G., Kreiman, U., Poggio, T.: A Theory of Object Recognition: Computations and Circuits in the Feedforward Path of the Ventral Stream in Primate Visual Cortex. Technical report, Massachusetts Institute of Technology Computer Science and Artificial Intelligence Laboratory, CBCL-259 (2005)
28. Short, T.L.: Darwin's concept of final cause: neither new nor trivial. Biology and Philosophy 17, 323–340 (2002)
29. Treisman, A.M., Gelade, G.: A feature-integration theory of attention. Cognitive Psychology 12(1), 97–136 (1980)
30. Ullman, S., Vidal-Naquet, M., Sali, E.: Visual features of intermediate complexity and their use in classification. Nature Neuroscience 5(7), 682–687 (2002)
31. Ungerleider, L.G., Haxby, J.V.: "What" and "where" in the human brain. Current Opinion in Neurobiology 4(2), 157–165 (1994)
32. Mishkin, M.M., Ungerleider, L.G., Macko, K.A.: Object vision and spatial vision: two cortical pathways. Trends in Neurosciences 6, 414–417 (1983)
33. Walther, D., Koch, C.: Modeling attention to salient proto-objects. Neural Networks 19(9), 1395–1407 (2006)
34. Wolfe, J.M., Horowitz, T.S.: What attributes guide the deployment of visual attention and how do they do it? Nat. Rev. Neurosci. 5(6), 495–501 (2004)

Optimizing a Conspicuous Point Detector for Camera Trajectory Estimation with Brain Programming

Daniel E. Hernández[1], Gustavo Olague[1,*], Eddie Clemente[1,2], and León Dozal[1]

[1] Proyecto EvoVisión,
Departamento de Ciencias de la Computación, División de Física Aplicada,
Centro de Investigación Científica y de Educación Superior de Ensenada,
Carretera Ensenada-Tijuana No. 3918, Zona Playitas, Ensenada, 22860, B.C., México
{dahernan,olague,eclemen,ldozal}@cicese.mx
http://cienciascomp.cicese.mx/evovision/
[2] Tecnológico de Estudios Superiores de Ecatepec. Avenida Tecnológico S/N, Esq.
Av. Carlos Hank González, Valle de Anáhuac, Ecatepec de Morelos

Abstract. The interaction between a visual system and its environment is an important research topic of purposive vision, seeking to establish a link between perception and action. When a robotic system implements vision as its main source of information from the environment, it must be selective with the perceived data. In order to fulfill the task at hand we must contrive a way of extracting data from the images that will help to achieve the system's goal; this selective process is what we call a visual behavior. In this paper, we present an automatic process for synthesizing visual behaviors through genetic programming, resulting in specialized prominent point detection algorithms to estimate the trajectory of a camera with a simultaneous localization and map building system. We present a real working system; the experiments were done with a robotic manipulator in a hand-eye configuration. The main idea of our work is to evolve a conspicuous point detector based on the concept of an artificial dorsal stream. We experimentally show that it is in fact possible to find conspicuous points in an image through a visual attention process, and that it is also possible to purposefully generate them through an evolutionary algorithm, seeking to solve a specific task.

Keywords: Evolutionary Visual Behavior, Multiobjective Evolution, Purposive Vision, SLAM, Conspicuous Point Detection.

1 Introduction

Active vision is a research area that is based on the relationship between perception and action. It studies how vision based systems manipulate their input visual information to solve a given task, as a result of the interaction between the vision system and the environment. For example, in active vision a robotic

---

* Corresponding author.

O. Schütze et al. (eds.), *EVOLVE - A Bridge between Probability, Set Oriented Numerics, and Evolutionary Computation III*, Studies in Computational Intelligence 500,
DOI: 10.1007/978-3-319-01460-9_6, © Springer International Publishing Switzerland 2014

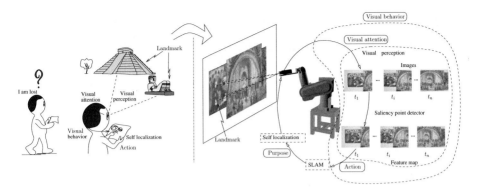

Fig. 1. A visual behavior is defined as the sequence of decisions that a system takes based on the information it perceives from its environment and how it performs the set of actions required to achieve its goal.

system is able to manipulate the attitude of a camera to achieve some task, or purpose, related to the observation of the environment where the robot acts [3]. A distinctive characteristic of active vision is the idea that the observer is capable of engaging into some kind of activity whose purpose is to change the visual parameters according to the environmental conditions [9]. In this way, purposive vision, as an information process, does not function as an isolated entity, but as a part of a bigger system that interacts with the world in highly specific ways ([1],[2]). The idea studied in this paper is that a purposive visual system is integrated into a more complex system whose interaction with the environment are done in a specific way opposed to a general manner [10]. Therefore, the aim of purposive or behavioral vision is to evolve a visual routine, via an evolutionary algorithm, whose overall goal is to adapt the visual program to a specific task. This work could be understood as part of the new research area of evolutionary computer vision, see [16].

Figure 1 illustrates the problem that we would like to approach using a camera mounted at the end of a robotic manipulator. The idea is that a visual behavior requires specific information related to the task that is being confronted. On the left-side of the figure, the visual behavior performed by a person is related to the extraction of visual information needed to read a map; as well as, the mental activity that is applied to extract the information needed to find an object within a scene. In this way, the person performs a set of actions including a visual behavior that needs to accomplish a number of goals, such as: visual perception and self-localization, in order find its way through an environment. On the right-side of the figure, we can see the necessary steps in a self-localization and map building (SLAM) process. This process is normally modelled as an estimation task, meaning, the system performs an action based on the information captured by the perception mechanism with the purpose of self-localization. In this system, the visual routine is done as a conspicuous point detector, and the action is executed by the SLAM method. In this way, the objective is to synthesize a specialized detector to estimate the trajectory of the camera over a specific path, evaluated through a SLAM system. Therefore, the aim of this paper is to experimentally show that it is possible to evolve a specialized visual routine, based on a visual attention model for the trajectory estimation of a camera mounted on a robot.

The remainder of this paper is organized as follows. First, we review the concept of visual behavior as a conspicuous point detection based on the artificial dorsal stream mode. Then, the simultaneous localization and map building system where the behavior functions, as well as the process for specializing the visual behavior through genetic programming are described. Thirdly, we describe the multiobjective evolutionary algorithm to synthesize conspicuous point detectors. Finally, we present results of a real working system followed by our conclusions and future work.

2   Visual Behavior

The main objective in this work is the synthesis of visual routines through genetic programming. The core functionality of the visual routine is a conspicuous point detector based on a visual attention model; this detector should be adapted to work within a SLAM system in order to estimate the trajectory of a camera. This section describes the three main aspects of our work: the concept of conspicuous point detection; followed by the SLAM system used as a testbed for the detectors; and finally, the optimization of the visual routines through an evolutionary process for conspicuous point detection.

2.1   Conspicuous Point Detection

The concept of conspicuous point detection used in this work was developed based on the theory behind an interest point detector. In their work, Trujillo and Olague showed that it is possible to evolve general purpose interest point detectors through genetic programming [19,14]. The goal in this work is to evolve specialized point detectors, or visual behaviors, but at the same time maintaining general purpose properties.

An interest point is a small region on an image which contains visually prominent information. The right side of Figure 1 depicts a robotic system in a hand-eye configuration that was used to evaluate the evolved interest point detectors. The idea is to create a visual behavior capable of extracting relevant visual information for solving a specific task; like the evaluation of the motion of the camera in a straight line with the pose estimation approach of a SLAM system. In particular, the value of importance of a pixel in an image is the result of a mapping $K : \mathbb{R}^+ \to \mathbb{R}$, this transformation is known as the *operator*, which should not be confused by the *detector*. The first one is applied to an image to obtain the importance of each pixel, and the latter is an algorithm that extracts all the interest points or regions in an image. In this way, most interest point detectors work as follows:

1. Apply the operator K to the input image in order to obtain the *interest image* I^*.
2. Perform a non-maximum suppression operation.
3. Establish a threshold to determine if a given maximum should be considered an interest point.

The idea in this work is to evolve a visual behavior, in such a way of maintaining general purpose characteristics, such as: *repeatability* and *dispersion*; while adapting the visual processing system to estimate a specific trajectory. In summary, the following properties are the desired characteristics for the evolved detectors:

1. *Repeatability*, the expected interest points should be robust to environmental conditions and geometric transformations.
2. *Dispersion*, the detector should be able to find points over the whole image.
3. *Trajectory estimation*, the detector must find useful information that simplifies the camera pose estimation computed through an SLAM system.

Hence, starting from the concept of an interest point detector; this work proposes to change the manner in which the importance of a pixel is evaluated by using the concept of visual attention. This is based on the model of an artificial dorsal stream instead of a single image operation. Thus, the proposed detector, called conspicuous point detector, works as follows:

1. Apply an evolved artificial dorsal stream algorithm to obtain a saliency map to be used as an interest image $I*$
2. Perform a non-maximum suppression operation.
3. Establish a threshold to determine if a given maximum should be considered as a conspicuous point.

To summarize, the resulting detected points by the proposed algorithm are similar to those found through an interest point detector, they are all prominent image regions; but since the process to detect the points is different, and it is inspired in the artificial dorsal stream model, we decided to name them conspicuous points.

Artificial Dorsal Stream Model. The dorsal stream, also known as the "where" pathway, is related to the visual processing necessary to determine the spatial location of the objects in the environment. Nevertheless, the description of the processes where the dorsal stream is involved is controversial. It is said to be involved in the guidance of actions, as well as, the localization of objects in space. The dorsal stream starts at the retina, from which it receives its main, if not all, input through magnocellular retinocortical layer of the lateral geniculate nucleus of the thalamus, and it projects into layer V1, which is part of the primary visual cortex, also called striate cortex, that is located at the back of the brain. Then, the dorsal stream goes through the V2 and V3 layers, which are a part of the brain known as extrastriate visual cortex; afterwards, the stream continues through the middle temporal area MT, and the medial superior temporal area MST; and finishing

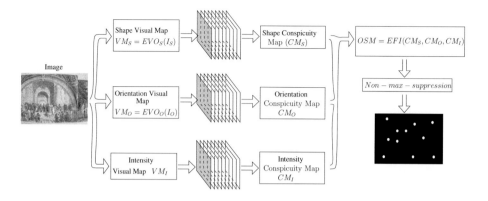

Fig. 2. The model of an artificial dorsal stream applied through a conspicuous point detection algorithm. It can be divided into two stages, the feature detection, where it seeks prominent features in three dimensions, and the feature integration, where it combines everything into a single map.

in the posterior parietal cortex and the adjacent areas. In general, it is granted that visual attention is performed by the dorsal stream, and the most accepted paradigm for visual attention, is the one presented in [11] known as feature integration theory. But when it comes to computational models, the first one was presented by Koch and Ullman [12]. They proposed to decompose the input image into different dimension in order to obtain a set of conspicuity maps and then integrate them into a saliency map. This is how we propose to find prominent points in an image; using the image decomposition into several dimensions enables the system to find attractive points based in different features, such as orientation, shape and intensity. Then, the idea is to find a plausible way of combining these different maps in order to obtain a single saliency map, which can be used as the interest image by the point detection algorithm.

The process of computing the interest image for the conspicuous point detection is biologically inspired, starting from the model of visual attention presented by Treisman and Gelade [18], known as an *Artificial Dorsal Stream* (ADS). The input for an ADS is an image and its output is an *Optimized Salient Map* (OSM). An OSM is an image whose pixel values represent the saliency of a point along the considered dimensions. In this way, the ADS is divided into two main stages: feature acquisition and feature integration; these stages can be seen in Figure 2. In the first phase, the input image is processed at three different and independent dimensions: intensity, orientation and shape. A set of visual operators is applied to the image creating one *Visual Map* (VM) per dimension. These maps represent the prominence of each pixel according to the corresponding dimension. For the second stage, the system combines the resulting VMs into a single map. Therefore, the output of the ADS is an OSM that expresses the prominence of each of the pixels in the input image. The following subsections describe, with more detail, the two stages along with the transformations that the image undergoes in each of them.

Table 1. Functions and terminals used to build the population for orientation

$$F_O = \{+, |+|, -, |-|, |I_{T_O}|, \times, \div, I_{T_O}^2, \sqrt{I_{T_O}},$$
$$log_2(I_{T_O}), \frac{I_{T_O}}{2}, D_x, D_y, G_{\sigma=1}, G_{\sigma=2}\}$$
$$T_O = \{I, D_x(I), D_{xx}(I), D_{xy}(I), D_{yy}(I), D_y(I),$$
$$G_{\sigma=1}(I), G_{\sigma=2}(I)\}$$

Feature Acquisition Stage. As mentioned above, this stage is composed of a set of visual operators that seek to highlight the pixels' prominence in three independent dimensions: intensity, orientation and shape. In their work Itti and Kosh [11] define the operators for these dimensions with a data driven approach. For the present system, a set of functions is developed through an evolutionary process to match the functionality of the dorsal stream in the natural visual system. This leads to a set of *Evolutionary Visual Operators* (EVOs), that will now be described. Note that, for the intensity dimension the system uses the input image since it is already an intensity map.

Orientation. The feature orientation detection function $EVO_O : I \to VM_O$; seeks to highlight prominent edges in the image. The values on the VM_O is the prominence of a certain pixel according to the operator EVO_O. Table 1 shows the set of functions and terminals used by the evolutionary algorithm to build the operators for the orientation dimension, where I is the input image, I_{T_O} are the elements in the terminal set T_O; or the output elements of the function set F_O; D_u represents the image derivatives in the direction $u \in \{x, y.xy, xx, yy\}$, and G_σ corresponds to a Gaussian filtering with the given σ value.

Shape. For the shape dimension, the system uses the function $EVO_S : I \to VM_S$; the aim of this operator is to accentuate interesting points based on the appearance and structure of the objects in the image, through mathematical morphology. The input nodes used to build this operator can be seen in Table 2.

Generation of Conspicuity Maps. After generating the three VMs, one per feature dimension, by using the aforementioned operators and the input image

Table 2. Functions and terminals used to build the population

$$F_S = \{+, -, \times, \div, round(I_{T_S}), floor(I_{T_S}),$$
$$ceil(I_{T_S}), dilation_{diamond}(I_{T_S}), dilation_{square}(I_{T_S}),$$
$$dilation_{disk}(I_{T_S}), erosion_{diamond}(I_{T_S}),$$
$$erosion_{square}(I_{T_S}), erosion_{disk}(I_{T_S}), skeleton(I_{T_S})$$
$$boundary(I_{T_S}), hit - miss_{diamond}(I_{T_S}),$$
$$hit - miss_{square}(I_{T_S}), hit - miss_{disk}(I_{T_S}),$$
$$top - hat(I_{T_S}), bottom - hat(I_{T_S}), open(I_{T_S}),$$
$$close(I_{T_S})\}$$
$$T_S = \{I\}$$

for the intensity dimension, the next step is to obtain the conspicuous maps (CMs) through the center-surround function $CS : VM \rightarrow CM$, which tries to emulate the center-surround receptive fields found in the natural dorsal stream. In order to achieve this functionality, the system creates a prism $VM_l(\alpha)$ with nine levels, each at a different spatial scale $\alpha = \{1, 2, ..., 9\}$. In this way, an across-scale subtraction $\ominus$ is performed leading to a set of center-surround maps $VM_l(\omega)$ for which the value of a pixel increases according to the contrast along its neighbors at different scales $\omega = \{1, 2, 3, 4, 5, 6\}$. Finally, the $VM_l(\omega)$ maps are merged using an across-scale addition $\oplus$ in order to obtain a conspicuous map CM_l per feature dimension. Thereafter, the resulting CMs must be combined using the feature integration operator in order to generate a single saliency map. This procedure is explained in the following section.

Feature Integration for a Saliency Map. Here we describe how the system combines the resulting CMs aiming to create a single map. We identify this operation as the *Feature Integration*, defined as follows: $RFI : CM_l \rightarrow OSM$, where CM_l are the conspicuous maps for the three feature dimensions $l = \{O, S, I\}$. Today, there is a lack of knowledge offering an explicit description of this process in the natural system; since, it is uncertain how the brain makes the CMs integration and in which region the saliency map is conceived. But as it has been stated, in this work, the visual routine was generated in a purposive manner. Therefore, a saliency map is built using an *Evolved Feature Integration* (EFI) function, leaving the task of establishing a good way to combine the maps to the artificial evolutionary algorithm. The set of functions F_{fi} and terminals T_{fi} used by the evolutionary algorithm to obtain this operator can be found in Table 3.

Table 3. Functions and terminals used to build the population

$$F_{fi} = \{+, |+|, -, |-|, |I_{T_{fi}}|, \times, \div, I^2_{T_{fi}}, \sqrt{I_{T_{fi}}},$$
$$log_2(I_{T_{fi}}), D_x, D_y, G_{\sigma=1}, G_{\sigma=2}\}$$
$$T_{fi} = \{CM_O, CM_I, CM_S, D_x(CM_O), D_{xx}(CM_O),$$
$$D_{xy}(CM_O), D_{yy}(CM_O), D_y(CM_O), ...\}$$

The resulting OSM was contrived through the combination of the CMs, where the values in the map represent the prominence of each pixel. Therefore, we propose to use the OSM as an interest image for the feature detection algorithm; since, the operation is quite different from the typical operator used for interest point detection, we will refer to the points found with this methodology as Conspicuous Points (CPs).

3   Trajectory Estimation with a SLAM System

For this work, the task that we focus on is the camera trajectory estimation through a visual behavior. To accomplish this task, the conspicuous point detector (CPD) is integrated into a simultaneous localization and map building (SLAM)

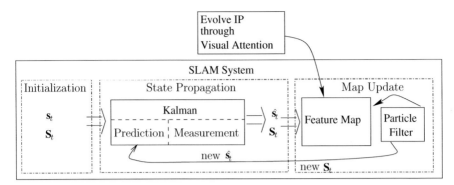

Fig. 3. Block diagram of the SLAM system; the proposed evolutionary algorithm for the specialization of the conspicuous point detectors is integrated in the map update stage

system. In this section, we describe how the system works and its relationship with the visual routine. The behavior must highlight useful information in the perceived images in order to be able to locate the camera over time. The SLAM system used to evaluate the efficiency of the visual behaviors in the corresponding task is based on the MonoSLAM system presented by Davison *et al.*, see [5]. The system uses a single camera, it implements a feature based map for the environment representation, and the localization process is approached as a state estimation task with a Kalman filtering algorithm. The concept of conspicuous points is applied in order to find prominent visual landmarks in the environment, which are then used to build the visual features map. The SLAM process is divided into three stages: initialization, state propagation, and map update.

Initialization Stage. In order to estimate the trajectory of the camera, the system has to determine its position along its movement; thus, the state of the SLAM system is defined by the camera's pose, position and orientation, and speed. In this manner, the process estimation of the camera's attitude $\hat{x}_v$, defined by its position r and orientation q, and linear and angular speeds, v and ω, is approached through Kalman filtering. Coupled with these variables, the system maintains an environment map, which is defined through a sparse set of visual landmarks. Therefore, the system tracks the position of the camera, along with the spacial position for each feature y_i composing the map. Hence, the system estimates the state $\mathbf{s}_t$, which contains the information about the camera and the feature map. The state estimation $\hat{s}_t$ is coupled with the uncertainty estimation $\mathbf{S}_t$. The initial state of the system is assumed to be known; in other words, the camera position is known along with a starting map, which consists of four landmarks on a calibration plane. The state $\mathbf{s}_t$ and its covariance $\mathbf{S}_t$ are defined as follows

$$
\mathbf{s}_t = \begin{pmatrix} \hat{x}_v \\ \hat{y}_1 \\ \hat{y}_2 \\ \vdots \end{pmatrix} \quad \mathbf{S}_t = \begin{bmatrix} S_{xx} & S_{xy_1} & S_{xy_2} & \cdots \\ S_{y_1x} & S_{y_1y_1} & S_{y_1y_2} & \cdots \\ S_{y_2x} & S_{y_2y_1} & S_{y_2y_2} & \cdots \\ \vdots & \vdots & & \end{bmatrix}.
$$

State Propagation. The state estimation is performed through a Kalman filtering method, and it is divided into two stages, prediction and measurement. The future states of the camera are assumed to follow a dynamic model of the form of

$$\mathbf{s}_t = \mathbf{A}s_{t-1} + \mathbf{w}_t,$$

where $\mathbf{A}$ is the state transition matrix, and $\mathbf{w}_t$ is the process noise. At each time step, the filter makes a *prediction* of the current state based on the previous state and the dynamic model. The prediction s_t^- is also known as the *a priori* state, together with its error matrix $\mathbf{S}_t^-$; this first step is called *prediction stage*. The dynamic model is defined by

$$s_t^- = \mathbf{A}s_{t-1},$$

$$\mathbf{S}_t^- = \mathbf{A}\mathbf{S}_{t-1}\mathbf{A}^T + \lambda_w,$$

where $\lambda_\mathbf{w}$ is the process covariance that models noise, assumed to be a white Gaussian noise.

Measurement Stage. The idea in this stage is to use the images captured by the camera to improve the state estimation. Hence, the measurements $\mathbf{z}_t$, represent the image location at time t of the visual landmarks, and are related to the current state of the camera, defined by

$$\mathbf{z}_t = \mathbf{C}\mathbf{s}_t + \mathbf{v}_t,$$

where $\mathbf{v}_t$ is the measurement noise, and $\mathbf{C}$ relates the camera pose to the images position of the landmarks. The measurement model is then used to establish an *a posteriori* state estimate $\hat{\mathbf{s}}_t$ together with its error matrix $\mathbf{S}_t$ by incorporating the measurements $\mathbf{z}_t$ through the following equation

$$\hat{\mathbf{s}}_t = \mathbf{s}_t^- + \mathbf{G}_t(\mathbf{z}_t - \mathbf{C}\mathbf{s}_t^-),$$

$$\mathbf{S}_t = \mathbf{S}_t^- - \mathbf{G}_t\mathbf{C}\mathbf{S}_t^-,$$

where $\mathbf{G}_t$ is known as the Kalman gain, defined as

$$\mathbf{G}_t = \mathbf{S}_t^- \mathbf{C}^T(\mathbf{C}\mathbf{S}_t^-\mathbf{C}^T + \lambda_v)^-1,$$

with λ_v being the measurement covariance. In this manner, the *a posteriori* state estimation at time t is complete; therefore, the state $\hat{\mathbf{s}}_t$ together with its uncertainty matrix $\mathbf{S}_t$ is ready for the next estimation cycle.

Map Update. Now that we have a good estimation of the camera's position, the system must capture more information of the environment in order to broaden its representation, this is done by extending the feature map. In this manner, the system applies the conspicuous point detection algorithm to find landmarks in the environment, which may enhance the state estimations in future time steps. In this way, the detector must bring, implicitly, the required characteristics on the conspicuous points to help the system solve the position estimation task.

The final step for the map update is to determine the spacial position of the visual landmarks. Since, the system uses monocular vision, it uses a particle filter to establish the depth of the visual landmarks. Once the depth is known for a certain point, it is added to the map for further estimations.

After these three steps, the estimation cycle of the Kalman filter is complete, and the position estimation is ready for the next iteration. This estimation cycle is executed for each image captured by the camera.

3.1 Evolving an ADS for Conspicuous Point Detection

This section describes the process of specialization of the ADS through the application of what we call brain programming (BP). This variation of genetic programming gets its name from the fact that it evolves not only a single function, but a set of them embedded into a hierarchical structure and coupled with some pre-designed procedures, aiming to emulate the functionality of an organ, such as the brain. For this work, the proposed genotype is formed by three operators, which can be seen as the genes, and when they are put together within a particular structure (program) they form a complex chromosome. The corresponding phenotype for the proposed model can be seen in Figure 2. Therefore, the algorithm in charge of processing the visual information is able to extract the conspicuous points within the image, by mimicking the functionality of an ADS, which is a single processing module. The aim of the proposed representation is to encourage the construction of complex functions capable of solving the task at hand; in this case, the construction of a conspicuous point detector to be applied inside the SLAM system previously described. Then, it is important to note that each individual of the population is understood as a complete ADS adapted to the purpose of extracting conspicuous points from the image, each individual is defined by the operators merged within an specific structure, and it is therefore not just a list of tree-based programs but rather an information processing system as shown in Figure 2. The evolved detectors are tested within the SLAM system described in Section 3. For this particular work, the testing trajectory of the detectors is the camera moving on a straight line, parallel to a wall rich in information. The SLAM system estimates the camera position during its trajectory using the evolved detector within the feature map building stage of the SLAM. Then, the fitness of the detector is evaluated through the quality of the position estimations, as well as, some other qualities of the extracted points. The evolved detectors should exhibit the properties mentioned in Section 2.1. Due to the amount of work required for the evolutionary algorithm, a camera motion sequence was captured in order to be used by the off-line evaluation to compute the detectors. The hypothesis for this work is that a good detector for the camera's trajectory estimation is one with high repeatability and high point dispersion. Which should bring stable landmarks and a disperse feature map. The following functions are used to evaluate the evolved detectors and are defined from a minimization perspective:

Repeatability. The average repeatability $r_K(\epsilon)$ is calculated for the operator K by evaluating the repeatability between two consecutive images, using a neighborhood of size ϵ. It is important to note that the repeatability is calculated using the position of the camera produced by the highly-accurate robot movement instead of using the homographies between the images.

$$rI_i(\varepsilon) = \frac{|R_{I_i}(\varepsilon)|}{min(\gamma_{i-1}, \gamma_i)},$$

where $\gamma_{i-1} = \left|\{x^c_{i-1}\}\right|$ and $\gamma_i = \left|\{x^c_i\}\right|$ are the number of points detected in images I_{i-1} and I_i. $R_{I_i}(\varepsilon)$ represents the set of pairs of points (x^c_{i-1}, x^c_i) that were found in two consecutive images within a region of radius ε:

$$f_1 = \frac{1}{r_K(\epsilon) + c_1},$$

where c_1 is a constant to avoid an invalid division.

Dispersion. To evaluate the detectors, $\mathcal{D}_p(K)$ is the average dispersion of the extracted points from the image sequence using the set individual K; where $c_2 = 10$ is a normalization constant.

$$f_2 = \frac{1}{e^{\mathcal{D}_p(K)-c_2}}.$$

The point dispersion in image I_i is calculated using the points' entropy $D(I, X) = -\sum P_j \cdot log_2(P_j)$ where X is the set of detected points and P_j is approximated using a histogram.

Trajectory Adjustment. The fitness of the individuals for calculating the camera motion is measured using the mean squared error of the estimated trajectory using the real straight-line trajectory given by the high-accurate robotic arm.

$$f_3 = \sum_{i=1}^{M} \frac{[x_{v_i} - \hat{x}_{v_i}]^2}{M}$$

Multiobjective Visual Behavior Evolution. For our work, we took a multiobjective (MO) approach which allows us to incorporate several optimization criteria to evaluate the evolved detectors, see [15]. It is important to note that in a MO algorithm the result is a set of Pareto optimal solutions, rather than a single best individual, see [7]. In our algorithm, each individual represents a conspicuous point detection program, which is composed of three operators embedded within a complex program, one used to generate the orientation VM_O, the other for the shape VM_S, and the last one for the feature integration process. These operators are created using the sets of functions and terminals defined in Section 2.1. The evolutionary process was executed using a Unibrain Fire-I camera mounted on a Stäubli RX-60 robot with six degrees of freedom. The camera

Table 4. Parameters used in the multiobjective GP for the synthesis of conspicuous point detectors

Parameters	Description
Population	30 individuals
Generations	30 iterations
Initial population	Ramped Half-and-Half
Genetic operations probabilities	Crossover $p_c = 0.85$
	Mutation $p_\mu = 0.15$
Max three depth	6 levels
File size (SPEA2)	15
Parent selection	15

captured monochromatic images with a resolution of 320×240 pixels. The testing trajectory was the robot moving in a straight-line, parallel to a wall rich in visual information, the length of the path is 70 centimeters, and it was represented by a set of 150 images. The parameters for the evolutionary algorithm can be found in Table 4. In this manner, each individual is evaluated through its functionality within the SLAM system. The image sequence is also employed to evaluate the repeatability and dispersion of each of the individuals. Next the genetic operations are executed according to the SPEA2 [20] algorithm for parent selection, one-point crossover and sub-tree mutation acting independently on each of the three operators in order to create the new offspring.

Therefore, the evolutionary algorithm implemented for our work is hierarchical, the operations do not have a real meaning by them self, since they are adapted to work together within a specific structure. It is also said to be heterogeneous, as mention earlier, we implemented different sets of functions and operators per each desired EVO. Finally, as described in this section, it is multiobjective, due to the aiming of the adaption of multiple detectors to the given task by evaluating their performance from three perspectives: repeatability, dispersion, and trajectory estimation.

4   Experimental Results

This section describes the results from the evolutionary process, in order to outline the implemented system that synthesizes an artificial dorsal stream using a conspicuous point detection process. The resulting population can be seen in Figure 4, which shows the distribution of the individuals along the fitness space. Also, in Figure 4 we can see an aerial view from the Repeatability-Dispersion plane, which shows the individuals group in the corner near the origin of the plane. The meaning can be interpreted as representing the set of individuals that achieve high performance along these two fitness functions creating point detectors that find highly repeatable but dispersed points. In Figure 4, we can see lateral views from the MSE-Repeatability and MSE-Dispersion planes. Moreover, we can observe that the individuals are distributed along the MSE axis, meaning it is harder to find fitter solutions with respect to this evaluation function,

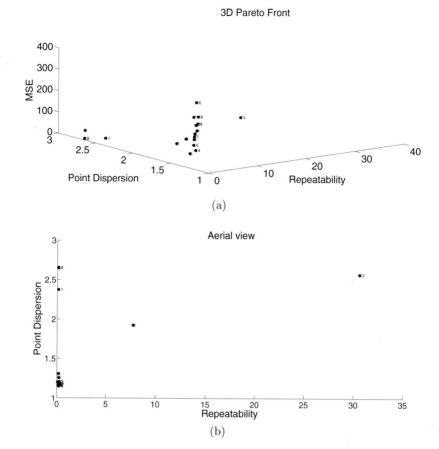

Fig. 4. Resulting Pareto front for the evolutionary process along the Repeatability-Dispersion plane

which results natural, since this is the function related to the task of tracking the camera. This distribution also means that this fitness function is in conflict with the other two. The Table 5 lists the non-dominated individuals forming the Pareto front produced with the system just described. The resulting operators are called conspicuous point detectors for SLAM (CPSLAM). These individuals were tested inside a monocular vision SLAM system described in Section 3. All experiments were carried out in a Dell Precision T7500 workstation, with a Intel Xeon 8 Core processor, an NVIDIA Quadro FX 3800 video card running the OpenSUSE 11.1 linux as an operating system.

It is interesting to note that most of the individuals do not use all the dimensions for the construction of the saliency map, which can be related to the fact that the evaluation trajectory is very simple. The most applied dimension is the orientation, meaning that this should be considered as the most helpful

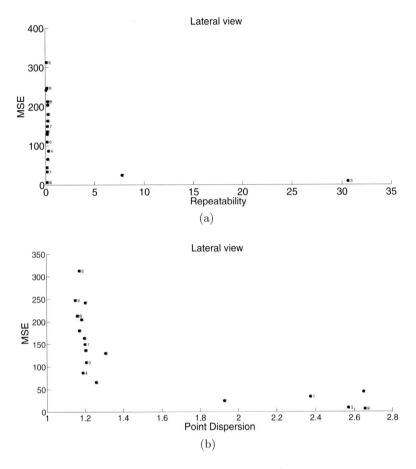

Fig. 5. Resulting Pareto front for the evolutionary process along the Dispersion-MSE plane

information for solving this particular straight line trajectory motion, due to the edge detection capabilities of the proposed operations.

Figure 6 shows the change on the diversity of the individuals in the population. Diversity is defined as the percentage of uniqueness of the solutions in the population. As mentioned earlier, there are three operator inside each individual, and we observe the normal tendency of an evolutionary algorithm. The population starts with high diversity, since it is created pseudo-randomly but it grows due to selection and crossover; nevertheless, we can also see some increment due to mutation. Since our evolutionary process is heterogeneous we evaluated the diversity for each operator separately. Measuring the diversity in our process is important because it shows that brain programming is capable of exploring different regions of the solution space, and it also helps to compare solutions that achieve similar results.

Table 5. Individuals found after the evolutionary process creating a Pareto front

Name	Operator	Fitness
CPSLAM1	$EVO_O = round(D_x(D_y(I))) + 0.09$ $EVO_S = hit - miss_{diamond}(I)$ $EFI = \dfrac{CM_I}{D_y(CM_S)}$	$f_1 = 0.124$ $f_2 = 2.647$ $f_3 = 44.51$
CPSLAM2	$EVO_O = thresh(log_2(\lfloor infimum(D_x(D_x(I)), I)\rfloor))$ $EVO_S = dilate_{square}(\frac{I}{0.93})$ $EFI = \mid \dfrac{(\lceil CM_I - D_x(D_y(CM_I))\rceil)}{0.63} \mid$	$f_1 = 7.774$ $f_2 = 1.926$ $f_3 = 24.459$
CPSLAM5	$EVO_O = G_{\sigma=1}(D_y(D_y(I)))$ $EVO_S = \lceil I \rceil$ $EFI = (D_x(D_y(CM_O)))^2$	$f_1 = 0.195$ $f_2 = 1.258$ $f_3 = 65.67$
CPSLAM11	$EVO_O = infimum(I^2, infimum((D_x(supremum(D_y(I), D_X(I)/2)))^2, TMP))$ $TMP = \mid D_x(D_x(I)) - G_{\sigma=1} * I \mid$ $EVO_S = kSust(\lfloor Erode_{diamond}(Dilate_{square}(I))\rfloor, 0.05)$ $EFI = round((D_y(D_y(CM_I)))/0.87)$	$f_1 = 0.134$ $f_2 = 2.373$ $f_3 = 33.67$
CPSLAM13	$EVO_O = \sqrt{thresh(supremum((D_x(D_x(I)))^{\frac{1}{0.85}}, round(I)))}$ $EVO_S = hit - miss_{diamond}(perimeter((I/0.97)) + (I + 0.55)^{\frac{1}{0.31}})$ $EFI = (round(D_y(D_y(CM_I)))/0.87)^2$	$f_1 = 30.592$ $f_2 = 2.569$ $f_3 = 9.164$
CPSLAM14	$EVO_O = \sqrt{\lceil \mid D_x(I) - \lceil D_y(D_y(I))\rceil \mid \rceil}$ $EVO_S = hit - miss_{diamond}(perimeter(I/0.97) + (I + 0.55)^{\frac{1}{0.31}})$ $EFI = \mid \mid e^{D_x(D_y(CM_S)) - CM_S} - (D_y(D_y(CM_I))/0.87) \mid \times G_{\sigma=2} * (D_x(D_x(CM_S)))$ $-D_x(D_x(CM_O))) + ((\mid CM_O + D_x(CM_I) \mid - G_{\sigma=1} * D_x(D_x(CM_S))) + 0.18) \mid$	$f_1 = 0.279$ $f_2 = 1.189$ $f_3 = 86.504$
CPSLAM18	$EVO_O = \sqrt{thresh(supremum((G_{\sigma=1} * D_y(D_y(I)))^{\frac{1}{0.85}}, round(G_{\sigma=1} * I)))}$ $EVO_S = hit - miss_{diamond}(\lceil I \rceil + (I + 0.55)^{\frac{1}{0.31}})$ $EFI = (CM_O + D_x(D_x(CM_O)) - G_{\sigma=1} * D_x(D_x(CM_S)) + 0.18$	$f_1 = 0.171$ $f_2 = 1.151$ $f_3 = 200.47$
CPSLAM19	$EVO_O = infimum(I^2, D_y(I))$ $EVO_S = hit - miss_{disk}(I)$ $EFI = round(D_y(D_y(CM_S)))^2$	$f_1 = 0.124$ $f_2 = 2.654$ $f_3 = 6.851$

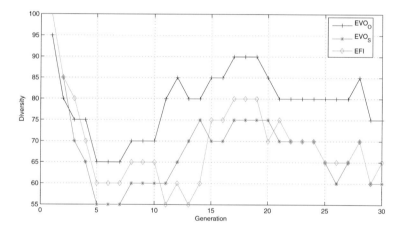

Fig. 6. Evolution of the diversity measure of the population over the evolution process

As mentioned previously, the evolutionary process applied in our work follows a multiobjective approach. Figures 7, 8 and 9 describe the evolution of the average fitness values of the population and the fitness values of the best individual at each generation for the repeatability, point dispersion and MSE of the conspicuous point detectors generated by each individual. In the case of the point dispersion and MSE fitness functions, we can see a regular tendency for a minimization process. They both tend to decrease, but for the repeatability value we can observe small increments; this is due to the fact that it is in conflict with the

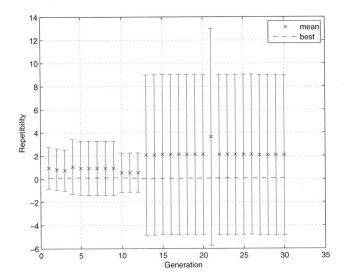

Fig. 7. Evolution of the Repeatability fitness, the graphic shows the best individual in each generation, and the population average

other fitness values. Note, that Figure 9 is presented in a logarithmic manner, this is due to the high values in the early steps of the evolutionary process; since, the population is created randomly and the detectors are not adapted for the task of solving the camera trajectory estimation task.

Figure 10 shows the execution of the system using the individual CPSLAM5 that achieves the better results. The sequence on the left correspond to the captured images where the ellipses on the images represent the computed visual landmarks. The right side of the figure depicts the error on the estimated trajectory along the XY and XZ planes.

In order to better understand the inner workings of a conspicuous point detection, Figure 11 shows how the resulting best individual $CPSLAM5$ decomposes an image in order to find prominent points. As we can see on the image, it first decomposes the image in three dimensions, but then, in the integration stage, it focuses on the orientation maps for creating the interest image. Finally, after the non-maximum suppression operation we can observe that the detector finds small areas of prominent points dispersed along the image, which is the kind of point distribution that is helpful for the SLAM system, since it has to create the sparse feature map for the position estimation task. Surprisingly, the final best result is similar to a typical interest point detector like those designed in previous research, see [14]. This result helps us show that brain programming is a viable technique for real world applications, since it is coherent with previous knowledge, and in a simple task, like the one approached in this chapter, brain programming is capable of finding human competitive solutions.

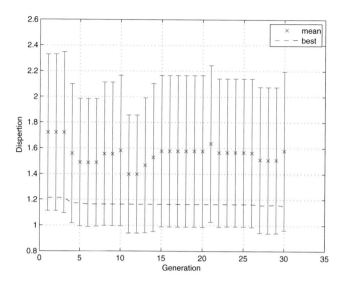

Fig. 8. Evolution of the Dispersion fitness value; best individual in each generation, and the population average

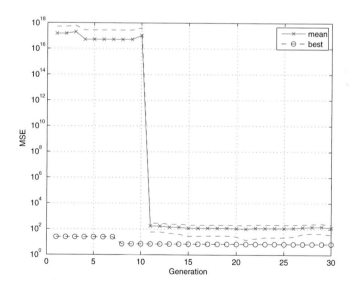

Fig. 9. Logarithmic view of the evolution of the MSE fitness; best individual in each generation, and the population average. Due to the large values of this fitness we present it in a logarithmic scale.

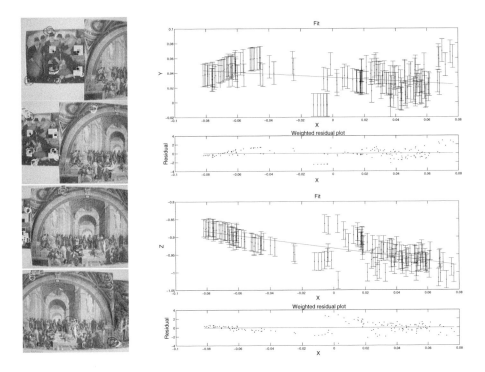

Fig. 10. System execution using the $CPSLAM5$ individual. On the left side we plot some images capture by the camera during its trajectory motion together with some points highlighted by the system. The right size shows the position estimation provided by the SLAM system.

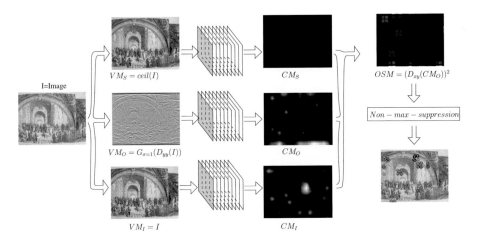

Fig. 11. Transformation of an input image performed by the $CPSLAM5$ detector

5   Conclusions and Future Work

The main objective of this work was to show that it is possible to purposefully generate visual behaviors through a multiobjective evolutionary algorithm, along with the idea of applying the concept of an artificial dorsal stream for a prominent point detection system, resulting in what we called conspicuous point detectors. Another important aspect of our work is that we focus on creating detectors for a specific task, in this case the trajectory estimation of a camera, rather than focusing on generic detectors. The results in this work show, that it is possible to perceive artificial evolution as a purposive process; also, that it is possible to find conspicuous points in an image through the concept of visual attention; and finally, that is it possible to design such detectors to solve a given task. Also, we can see on the results that exists in fact a conflict between the properties we seek in a conspicuous point detector, which shows that it is coherent to approach this as a multiobjective optimization process. The proposed perspective of behavior vision brings a rich research line of work, we list some of these next:

1. The synthesized detectors where generated using a simple straight line motion trajectory. It would be interesting to test the resulting detectors in more complex trajectories, and evolve the detectors using these new motions, and compare the resulting detectors.
2. Another problem is to apply the evolved conspicuous point detectors in other visual tasks in order to compare its efficiency against interest point detectors.
3. Also, we can compare the performance of a conspicuous point detector against other interest point detectors using a standard benchmark.
4. Moreover, another goal could be to introduce motion related visual operations to see if they improve the functionality of the detectors.
5. Finally, the idea is to perform some tests in order to find which feature dimensions are better based on the environment where the system works.

Acknowledgements. This research was founded by CONACyT through the Project 155045 - "Evolución de Cerebros Artificiales en Visión por Computadora". First author supported by scholarship 267339/220773 from CONACyT. This research was also supported by TESE through the project DIMI-MCIM-004/08.

References

1. Aloimonos, J., Weiss, I., Bandyopadhyay, A.: Active vision. In: Proceedings of the First International Conference on Computer Vision, pp. 35–54 (1987)
2. Aloimonos, Y.: Active Perception, 292 pages. Lawrence Erlbaum Associates, Publishers (1993)
3. Ballard, D.: Animate Vision. Artificial Intelligence Journal 48, 57–86 (1991)

4. Clemente, E., Olague, G., Dozal, L., Mancilla, M.: Object Recognition with an Optimized Ventral Stream Model using Genetic Programming. In: Di Chio, C., et al. (eds.) EvoApplications 2012. LNCS, vol. 7248, pp. 315–325. Springer, Heidelberg (2012)
5. Davison, A.J.: Real-Time Simultaneous Localisation and Mapping with a Single Camera. In: Proceedings of the Ninth IEEE International Conference on Computer Vision, vol. 2, pp. 1403–1410. IEEE Computer Society, Washington, DC (2003)
6. Dozal, L., Olague, G., Clemente, E., Sánchez, M.: Evolving Visual Attention Programs through EVO Features. In: Di Chio, C., et al. (eds.) EvoApplications 2012. LNCS, vol. 7248, pp. 326–335. Springer, Heidelberg (2012)
7. Dunn, E., Olague, G.: Multi-objective sensor planning for efficient and accurate object reconstruction. In: Raidl, G.R., et al. (eds.) EvoWorkshops 2004. LNCS, vol. 3005, pp. 312–321. Springer, Heidelberg (2004)
8. Dunn, E., Olague, G.: Pareto Optimal Camera Placement for Automated Visual Inspection. In: IEEE/RSJ International Conference on Intelligent Robots and Systems, pp. 3821–3826 (2005)
9. Fermüller, C., Aloimonos, Y.: The Synthesis of Vision and Action. In: Landy, et al. (eds.) Exploratory Vision: The Active Eye, ch. 9, pp. 205–240. Springer (1995)
10. Hernández, D., Olague, G., Clemente, E., Dozal, L.: Evolutionary Purposive or Behavioral Vision for Camera Trajectory Estimation. In: Di Chio, C., et al. (eds.) EvoApplications 2012. LNCS, vol. 7248, pp. 336–345. Springer, Heidelberg (2012)
11. Itti, L., Koch, C.: Computational modelling of visual attention. Nature Review Neuroscience 2(3), 194–203 (2001)
12. Koch, C., Ullman, S.: Shifts in selective visual attention: towards the underlying neural circuitry. Hum. Neurobiol. 4(4), 219–227 (1985)
13. Lepetit, V., Fua, P.: Monocular Model-Based 3D Tracking of Rigid Objects: A Survey. Foundations and Trends in Computer Graphics and Vision 1, 1–89 (2005)
14. Olague, G., Trujillo, L.: Evolutionary-computer-assisted design of image operators that detect interest points using genetic programming. Image and Vision Computing 29(7), 484–498 (2011)
15. Olague, G., Trujillo, L.: Interest Point Detection through Multiobjective Genetic Programming. Applied Soft Computing 12(8), 2566–2582 (2012)
16. Olague, G.: Evolutionary Computer Vision – The First Footprints. Springer (to appear)
17. Shi, J., Tomasi, C.: Good features to track. In: Proceedings of Computer Vision and Pattern Recognition, pp. 593–600 (1994)
18. Treisman, A.M., Gelade, G.: A feature-integration theory of attention. Cognitive Psychology 12(1), 97–136 (1980)
19. Trujillo, L., Olague, G.: Automated Design of Image Operators that Detect Interest Points. Evolutionary Computation 16(4), 483–507 (2008)
20. Zitzler, E., Laumanns, M., Thiele, L.: SPEA2: Improving the strength Pareto evolutionary algorithm. Technical report, Evolutionary Methods for Design (2001)

Optimizing an Artificial Dorsal Stream on Purpose for Visual Attention

Gustavo Olague[1,*], León Dozal[1], Eddie Clemente[1,3], and Arturo Ocampo[2]

[1] CICESE, Carretera Ensenada-Tijuana No. 3918,
Zona Playitas, Ensenada, B.C., México
{ldozal,olague,eclemen}@cicese.mx
[2] Facultad de Estudios Superiores Aragón, UNAM. Av. Rancho Seco s/n. Col.
Impulsora, Nezahualcoyotl. Edo. Mex., México
aoa@unam.mx
[3] Tecnológico de Estudios Superiores de Ecatepec, Avenida Tecnológico S/N, Esq.
Av. Carlos Hank González, Valle de Anáhuac, Ecatepec de Morelos, México

Abstract. Visual attention is a natural process performed by the brain, specifically by the dorsal stream, whose functionality is to perceive salient visual features. This chapter is devoted to the task of evolving an artificial dorsal stream (ADS) using the brain programming strategy. The idea is to state the problem of visual attention, normally studied as two parts: bottom-up and top-down, in terms of a unified approach following a teleological framework. Indeed, in this work visual attention is explained as a single mechanism that adapts itself according to a given task. In this way, brain programming is used to design ADSs. Experimental results show that this new approach can contrive ADSs useful in the solution of "top-down and bottom-up" visual attention problems. In particular, we present a solution to the size and missing pop-out problems that were unsolved previously in the literature.

Keywords: brain programing, visual attention, genetic programming.

1 Introduction

Visual attention is a skill, which allows to a creature, living or artificial, to direct their gaze rapidly towards the objects of interest in the visual environment [14]. The objects of interest refers to those objects or regions in the environment, which contain important information at a given time. Moreover, the visual attention mechanism is one of the most important mechanisms in the visual system because the brain is unable to process all visual information acquired along the entire visual field. In this way, there are two basic processes that define the problem of visual attention. The first basic phenomenon is due to the limited capacity of information processing. Therefore, only a small amount of information available to the retina can be processed and used to control a specific behavior.

---

* Corresponding author.

O. Schütze et al. (eds.), *EVOLVE - A Bridge between Probability, Set Oriented Numerics, and Evolutionary Computation III,* Studies in Computational Intelligence 500,
DOI: 10.1007/978-3-319-01460-9_7, © Springer International Publishing Switzerland 2014

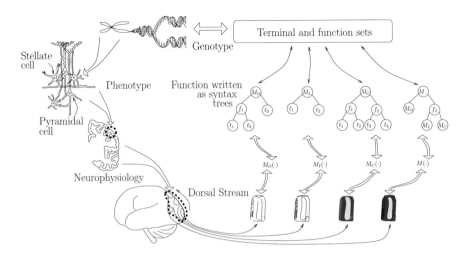

Fig. 1. This figure illustrates the analogy between the natural and artificial systems. The idea is based on replicating the functionality of a set of artificial tissues that conform what we called the brain programming (BP) system.

The second basic phenomenon is selectivity; in other words, the ability to filter unwanted information [7]. The brain can be extremely complex and despite rapid scientific progress much about how the brain works remains a mystery. In nature, there is a large diversity of brain anatomies that are characterized by the specialization of visual systems. Such diversity shows the power of evolution through adaptation. In this way, it has been argued that the evolution of specific visual mechanisms in the primate brain is the product of natural selection [4]. Contrary, in the recent past it was widely believed that human observers constructed a complete representation of everything in their visual field [8,25]. This has been amply refuted by a large amount of vision research.

In this work, we follow the idea that visual attention is controlled by both cognitive, or top-down (TD) factors, such as knowledge, expectation, and current goals; as well as, reactive stimulus, or bottom-up (BU) factors, that refers to sensory stimulation like novelty, unexpectedness, and brightness, see [6]. Moreover, the low level mechanisms for feature extraction act in parallel over the entire visual field using the TD and BU systems in order to provide the signs that highlight the image regions. Afterwards, attention is focused sequentially on the highlighted regions of the image in order to make a posterior analysis, or simply by processing them [26,14].

1.1 Problem Statement

Since the late 19th century visual attention has been studied by researchers from different scientific disciplines; such as: neurologists, physiologists, psychologists, and in the last three decades by people working on computer vision. In particular, we believe that a new community of researchers working with genetic and evolutionary methods could be interested in the problem of visual attention, see [18]. Note that, all these researchers believe that it is necessary to implement this property within artificial systems since it is reasonable to assume such process as a feasible way to reduce the complexity of visual information processing. In this way, considering the selectivity of visual information, the answers to the questions: "what features should be selected?" and "when to use those features?" are not evident. Moreover, a problem arises after the feature detection stage known as feature combination. Combining different features, such as color, orientation and shape, within a single saliency representation becomes complex since these features came from different visual dimensions. The complexity increases when you are looking for a particular object, and it is necessary to filter the information to stress the features of the desired object. In this work, brain programming (BP) is used to address this problem. In this way, the idea is to apply BP as the mechanism to obtain the most suitable artificial dorsal streams (ADSs) programs that are capable of pursuing the desired goals.

2   Visual Attention Processing

This section proposes a new approach for visual attention with the aim of organizing the whole system as a single functional entity that changes its operation according to a purpose, but without changing its general structure. In this way, contrary to most traditional approaches that represent the visual attention model through the division of the process into reactive and volitive parts; our proposal provides the simplicity and uniqueness to endow a machine with the ability of designing visual attention programs that fit specific goals. Next, the main theory is reviewed in order to understand our approach to visual attention.

2.1   Classical Approach to Visual Attention

Visual attention functionality is related to the brain areas around the dorsal stream. Thus, the dorsal pathway is defined as projecting from V1 through V2, V3, middle temporal area (MT), medial superior temporal area (MST) and finally to the posterior parietal cortex, see [27]. Nevertheless, there is a lack of consensus about the specific brain areas (structure, and functionality), that conform the dorsal stream. For example, in another theory the dorsal stream is also known as the "how" stream [17]; while, in the work described in [3] the dorsal stream areas do not correspond to the literature.

Nowadays, classical explanations of visual attention are in agreement that the dorsal stream functionality can be influenced by BU and TD factors. In this way, it is affirmed by [6] that there are two interacting neural systems involved in the control of BU an TD factors that control visual attention. As a result, the dichotomy of visual attention has inspired several computational models that are commonly based on only one of these two factors. For example, the research in computational neuroscience has traditionally separated their study; as well as, the implementation of visual attention using a benchmark system for human-visual gaze estimation [21,5] or for the solution of object recognition tasks. Contrary to this line of research, we propose to study visual attention from a teleological standpoint as a way of unifying through this framework both factors with the intention of considering visual attention as a single mechanism.

Next, both BU and TD factors are reviewed in order to introduce our approach with the aim of understanding the structure and functionality of visual attention. In this way, both factors should be studied through a unique process that is capable of adapting itself according to the pursued goal or goals.

"Bottom-Up" Control for Visual Attention. In the literature the idea of BU visual attention is related with involuntary attention, which is usually compared with the concept of a spotlight. This metaphor has been used by Posner *et al.* [22] to explain that visual attention operates "as a spotlight which improves the detection of events in their proximity". Actually, one of the best and easiest ways of implementing a set of tests is to study BU attention in terms of visual search. Commonly, the exploratory task is studied experimentally using a set of images containing challenging visual stimuli that are presented to an observer. For each image there is an object called target that is different from the rest. Today, the existing computational models are mostly BU models based on the feature-integration theory [26], that we review next. The first biologically, neurologically, and plausible computational model for BU visual attention was proposed by Koch and Ullman [14]. Later, Milanese [16] proposed a visual attention system using mechanisms, inspired from biological processes, which were adopted by the research community to create a whole new trend in visual attention systems. Some of these processes are color opponencies such as: red-green and blue-yellow; as well as, the center-surround difference present in the receptive field of the cortical cells. One of the most well-known models is probably that of Itti *et al.* [12], which provided software that popularize these theoretical processes. In summary, this can be considered as a very detailed model that proposes simple solutions to complex issues. Later, another breakthrough was proposed by Rensink [24,23] who introduced the notion of proto-objects and the interpretation of the apparent blindness of observers to recognize dramatic changes within a scene. Finally, Walter and Koch [28], showed that the proposed model can enhance the task of object recognition through the application of the concept of proto-object for visual attention tasks.

"Top-Down" Control for Visual Attention. Today, there is an agreement that TD cues play a key role in the processing of visual information. In particular, it is known that during the TD visual attention there are numerous connections between higher and simpler information processing areas. In this way, it is said that voluntary attention takes more time, and effort to accomplish high-level tasks in comparison with involuntary attention. This is because the target shares with the distractors two or more features, which forces the observer to perform a scanning of the whole scene.

The TD visual phenomenon, just explained, is usually studied in psychophysics through the so-called "cuing experiments". This type of experiments consists in presenting a "cue" that guides the observer's attention toward the target. In this way, it is said that cues may indicate *where* is the target, like in the case of an arrow pointing towards the target, or by answering the question of *what* is the target by means of finding the similarities between a picture, or written description of the target, see [9]. Thus, there have been several attempts to implement models using TD models. For example, Oliva *et al.* [19] propose an attentional model that uses knowledge about the distribution of features over the image in order to select salient regions. Peters and Itti [21] proposed a combined model BU/TD, in which they measure the ability of the model to predict the saccades of people playing video games. In this way, they improved the prediction by a margin that doubles the performance obtained by the BU model. The TD part computes a feature vector describing the "gist" of the image with the positions of saccades obtained from real observers that are used to train the model. Finally, a feature vector is calculated to generate the saccades prediction map. Recently, Borji, *et al.* [5] follow the same line of research proposed by Peters and Itti, but the system is based on a different approach that determines the position of the saccades with respect to the observer by applying a set of robust classifiers.

Thus, from a computational modeling standpoint TD factors are not a trivial task; in other words, emotions and desires are difficult concepts to model within computer science. Nevertheless, a purpose should not be confused with a desire; when we refer to a purpose, we talk in terms of whether the goals are achieved or not; i.e., the goals are always associated with a metric. Here lies the importance of modeling TD and BU mechanisms in teleological terms.

2.2   A Unified Approach for Visual Attention

Aristotle defined the final cause or *telos* as that for which something is done, its purpose. He also distinguishes between the *telos* and desire, as well as the consciousness and intelligence. Therefore, according to Aristotle, an organism, like a seed, has a purpose just as a person, see [2]. Later, Kant [13] wrote, in the "Analytic of Teleological Judgment", that organisms must be regarded in teleological terms, and in the "Dialectic of Teleological Judgment", he attempts to reconcile this teleological conception of organisms with a mechanistic account of nature. Everything can be completely explained by causality, except the organisms. In fact, the understanding of terms such as: final cause, end, purpose, end for which, good for which, in Darwin's documents is in relation to his thesis of natural selection, see [15].

From our standpoint, we define attention as the result of a single mechanism that is designed to obey a general purpose. For example, the most primitive purpose for life could be survivorship. But, the achievement of survivorship depends on many other particular tasks; for example, prey hunting, mating, predator escape, etc. In this sense, visual attention is capable to adapt to the kind of goal associated to the current task of the organism. In order to accomplish such task, it is necessary to have a unique and general visual attention structure capable of performing, by some temporal readaptation, the necessary functions to perform such task. Furthermore, considering the fact that most of the tasks involved in the design of BU and TD factors are complex, we could say that the space of possible readaptations is at least very large and discrete. Therefore, we define visual attention as follows.

Definition 1 (Visual Attention). *Visual attention is a process that designs a relationship between the different properties of the scene, which are perceived through the visual system with the aim of selecting a particular aspect.*

For these reasons, we consider visual attention as a single computational structure that performs BU and TD processes. In consequence, in this work visual attention is studied within a unified framework in order to evolve visual attention programs (VAPs) that will be adapted to specific tasks.

3 Purposive Evolution for Visual Attention

Nowadays, from a biological perspective, it is well-known that the development of specific visual mechanisms in the primate brains is associated to evolution; specifically, this is linked to natural selection as it is explained in evolutionary theory. Moreover, the theory of evolution is not exempt of the concept of purpose and vice versa. Charles Darwin was the one who brought the concept of purpose into consideration. Note that Darwin uses the term *final cause* systematically in his writings as documented by Lennox [15]. On the other hand, Barton [4] explains the evolution of primates brains in terms of the specialization of visual mechanisms; such as visual attention. Thus, this section describes the general structure of attention, which is biologically inspired and will be evolved to suit different objectives. The resulting evolved programs will be known as artificial dorsal streams. Moreover, following the same direction of Treisman, the description of the general approach is divided into two main stages: acquisition and integration; see [26].

3.1 Acquisition of Early Visual Features

In previous works of artificial visual attention, the operators are established according to particular visual characteristics and the manner in what neuroscientists describe the knowledge about how they are obtained; an approach that some authors refer to as data driven. But, the brain can be very complex and despite rapid scientific progress, much about how it works remains a mystery. However, it is well known that vision is useful for accomplishing certain tasks. In this way, the ADS is a function driven approach that considers the biological visual process from the standpoint of its functionality, paying special attention to its aim and exploiting the knowledge about a given task and the intrinsic characteristics of the scene, to create complex programs based on functions, called visual operators. Moreover, it is widely recognized that the operation of

Table 1. Functions and terminals used by EVO_O to create the orientation visual map VM_O

$$F_O = \{+, \, -, \, \times, \, \div, \, |+|, \, |-|, \, \sqrt{I_{out}}, \, I_{out}^2, \, log_2(I_{out}),$$
$$G_{\sigma=1}, G_{\sigma=2}, |I_{out}|, \frac{I_{out}}{2}, \, D_x, \, D_y\}$$
$$T_O = \{I_r, \, I_g, \, I_b, \, I_c, \, I_m, \, I_y, \, I_k, \, I_h, \, I_s, \, I_v, \, G_{\sigma=1}(I_r),$$
$$G_{\sigma=2}(I_r), \, D_x(I_r), \, D_y(I_r), \, D_{xx}(I_r), \, D_{yy}(I_r),$$
$$D_{xy}(I_r), \, \dots\}$$

the visual cortex, specifically the dorsal stream, is a product of the evolutionary process. For these reasons, we propose to use evolutionary computation to obtain these artificial visual operators. In summary, this section explains how to use specialized evolved visual operators (EVOs) for the acquisition of visual dimensions such as color, orientation and shape. Next, the EVO features used within the ADS are defined.

Orientation. In previous works the characteristic of orientation for images was computed in gray scales. Thus, our work proposes to evolve the property of orientation along different color bands of the image. In this way, a rich set of information is generated since edges, corners, and other similar features could appear easily highlighted within the color bands. Therefore, the evolutionary approach evolves a function $EVO_O : I_{color} \to VM_O$ that cooperates with the ADS in order to accomplish the task. The resulting EVO_O operation is a visual map VM_O for which the pixel value represents the feature prominence; in such a way, that the larger the pixel value, the greater the orientation prominence of the feature. This computation is performed through a set of functions and terminals that are provided in Table 1. The notation that was used is as follows. I_{T_O} can be any of the terminals in T_O; as well as, the output of any of the functions in F_O; D_u symbolizes the image derivatives along direction $u \in \{x, y, xx, yy, xy\}$; G_σ are Gaussian smoothing filters with $\sigma = 1$ or 2.

Color. In biology, the color is encoded through photoreceptor cells known as cones, which are located in the retina. However, a special case is the yellow color which is not perceived in the cones but in the retinal ganglion cells. Then, the dorsal pathway is composed of several tissues V1, V2 and V4, whose cells respond to color features. In this work, the characteristics of color information will be used as the building blocks to construct the EVO_C by applying color opponencies and simple arithmetic operations between the different color bands in the corresponding color space. In the same way, as in EVO_O, the evolutionary process uses a set of functions and terminals provided in Table 2 to evolve the feature along the color dimension. The result is a visual map $EVO_C : I_{color} \to VM_C$ containing the color prominent features.

Table 2. Functions and terminals used by EVO_C to create the color visual map VM_C

$$F_C = \{+, \, -, \, \times, \, \div, \, |+|, \, |-|, \, \sqrt{I_{out}}, \, I_{out}^2,$$
$$log_2(I_{out}), \, Exp(I_{out}), \, Complement(I_{out}) \, \}$$
$$T_C = \{I_r, I_g, I_b, I_c, I_m, I_y, I_k, I_h, I_s, I_v, RG_{oppn},$$
$$YB_{oppn} \, \}$$

Table 3. Set of functions and terminals used by EVO_S to create the shape visual map VM_S

$$F_S = \{+, \, -, \, \times, \, \div, \, round(I_{out}), \, \lfloor I_{out} \rfloor, \, \lceil I_{out} \rceil,$$
$$dilation_{diamond}(I_{out}), \, dilation_{square}(I_{out}),$$
$$dilation_{disk}(I_{out}), \quad erosion_{diamond}(I_{out}),$$
$$erosion_{square}(I_{out}), \quad erosion_{disk}(I_{out}),$$
$$skeleton(I_{out}), \quad boundary(I_{out}), \quad hit \, -$$
$$miss_{diamond}(I_{out}), \, hit - miss_{square}(I_{out}),$$
$$hit \, - \, miss_{disk}(I_{out}), \quad top \, - \, hat(I_{out}),$$
$$bottom - hat(I_{out}), \, open(I_{out}), \, close(I_{out})$$
$$\}$$
$$T_S = \{I_r, I_g, I_b, I_c, I_m, I_y, I_k, I_h, I_s, I_v\}$$

Shape. As in previous dimensions, the evolutionary process uses a set of functions and terminals provided in Table 3 to characterize the shape information used in our proposed system. Note, that we propose to describe these features through mathematical morphology. The result is a visual map $EVO_S : I_{color} \rightarrow VM_S$ containing the shape prominent features. This part is evolved with the aim to provide the information about shape and structure of the object of interest within the image. We would like to remark that the application of this kind of morphological functions has not been applied in previous research studying the ventral and dorsal streams.

Intensity. Finally, to obtain the intensity of pixels in the image the model averages the red, green and blue values for each pixel. The result of this operation is a visual map VM_I in which the pixel represents the prominence over the intensity space. The VM_I formula is written as follows:

$$VM_I = \frac{I_R + I_G + I_B}{3} \quad ,$$

where I_R, I_G, and I_B are the red, green and blue bands respectively.

Computing the Conspicuity Maps. The conspicuity maps (CMs) are obtained by means of a center-surround function that is applied in order to simulate the center-surround receptive fields [28]. This natural structure allows the ganglion cells to measure the differences between firing rates in center (c) and

Table 4. Set of functions and terminals used by EFI to create the object saliency map OSM

$$F_{fi} = \{+, \, -, \, \times, \, \div, \, |+|, \, |-|, \, \sqrt{I_{out}}, \, I^2_{out},$$
$$Exp(out), \, G_{\sigma=1}, G_{\sigma=2}, \, |I_{out}|, \, D_x, \, D_y\}$$
$$T_{fi} = \{CM_I, CM_O, CM_C, D_x(CM_I), D_y(CM_I),$$
$$D_{xx}(CM_I), D_{yy}(CM_I), D_{xy}(CM_I), \ldots \}$$

surroundings (s) areas of ganglion cells. First, a pyramid VM^{α}_l of nine spatial scales $S = \{1, 2, ..., 9\}$ is created for each of the three resulting VMs. Afterwards, an across-scale subtraction $\ominus$ is performed, resulting in a pyramid of center-surround maps VM^{ω}_l for which the value of the pixel is augmented as long as the contrast of their neighbors at different scales is higher. Finally, the $VM_l(\omega)$ maps are added using an across-scale addition $\oplus$ in order to obtain the conspicuity maps CM_l. At this stage, we have one CM for each feature. The CMs were obtained as explained in the Walther and Koch model [28]. Finally, the CMs are combined to obtain a single saliency map as explained in the next section.

3.2   Feature-Integration for Visual Attention

The saliency map (SM) defines the place for the most prominent locations of the image; given the characteristics of intensity, orientation, color and shape. In other words, the objective of this stage is to decide where attention should be directed at any given time. In neuroscience, an exact description about how the brain makes this integration, or where is located the saliency map in the brain is unknown. In this work, the problem statement considers that the problem must be addressed regarding the task to be performed. In other words, since the task needs to accomplish a goal; then, the main criterion should be the one that guides the suitable combination of characteristics. In this way, genetic programming is very useful since it provides a methodology to address the problem. Therefore, we decided to evolve the integration of CMs through a function that we called Evolved Feature Integration (EFI). Once the integration of features is performed, we get an optimized saliency map (OSM) indicating the location of the most prominent regions within the original image, known as proto-object (P_t). The definition of the EFI function is as follows:

$$EFI : CM_l \rightarrow OSM \; ; \; l \in \{O, C, I\} \; .$$

The evolutionary method uses the set of functions and terminals, listed in Table 4, to create a fusion operator that highlights the features of the object of interest.

Hence, an OSM is characterized through a proto-object P_t or a sequence of proto-objects $\{P_1, P_2, \ldots, P_i, \ldots, P_t\}$, see [23]. These structures provide the local descriptions using the concept of proto-object or salient region of the OSM, which is attended at time t. In the next section, we explain the evolutionary process used to obtain the ADS.

4   Brain Programming

In this section, we describe the main aspects for the evolution of ADSs using the brain programming (BP) strategy. In BP the chromosome is composed of several genes and each one is represented with a tree structure. At the chromosome level the whole genotype is described by the parallel set of functions acting over the orientation, color, and shape dimensions. While at the gene level the genetic operations are performed like in classical genetic programming. The design of BP embody an organic motivation in a sense of describing an organ or tissue, as a part of a living organism, and their complexity. We introduce a set of new concepts in order to deal with the evolution of complex structures, which are explained next.

The first phase of the BP is the training. In this phase the BP learns to focus a prominent object using an image database for training. The Algorithm 1 lists the steps that the BP performs in order to obtain the ADSs. In this work, we propose an ADS genotype that is robust because it is capable of encoding in a better way the phenotype of the dorsal stream. More specifically, the genotype consists of four trees; where each has a different and specialized functionality. Hence, each tree has its own independent set of functions and terminals. Unlike the classical GP that only works with a representation of a single tree using a unique set of functions and terminals. Thus, BP encodes the four tree-based programs within a hierarchical structure that defines the visual attention program. In this way, the functions and terminal sets of the four trees are listed in Tables 1, 2, 3, and 4 considering orientation, color, shape, and feature integration, respectively. The ADS genotype is created as follows: the first tree is an EVO_O; the second tree is called an EVO_C, the third tree is an EVO_S, and the fourth tree is named an EFI; each one with a maximum depth of 9 levels. The first one encodes the orientation, or the operation defining the orientation-sensitive cells in V1 [11]. The second one encodes the color, or the operation of photoreceptor cells and color-sensitive cells present in the layers V1 and V4 of the visual cortex. The third one models the shape feature that characterizes shape-sensitive cells present in layers V2 and V4 of the brain. Finally, the fourth one encodes the way in which the features are combined to obtain the saliency map, or operation of the posterior parietal cortex, see [10]. The algorithm initializes a population of 30 ADSs with a ramped half-and-half technique.

After initialization, the BP needs a well-posed fitness function. In this work, we propose to use the F-measure as the fitness function working in a machine learning framework in order to compare and select among several ADSs. This measure has already been used in previous works as evaluation for applications related to computer vision such as [20] and [1]. The calculation of the fitness of the ADS is the result of a comparison between a P_t attended by the ADS, and a manual location previously defined. Thus, attention and segmentation of an object, should follow an ideal visual attention criterion. In this way, a target is considered attended if a not empty subset of pixels that conform the object intersects the proto-object P_t. Moreover, another difference arises considering the existence of two-level complexity in the structure of the genotype. First,

---

Algorithm 1. Brain Programming Algorithm

Randomly create an initial population of *ADSs*.

repeat

  Execute each *ADS*, using the training image database, and compute its fitness.

  Select one or two *ADSs* from the population with a probability proportional to their fitness to participate in genetic recombination.

  Create a new *ADS* by applying genetic operations with specific probabilities.

  until An acceptable solution is found or some other stopping condition is met (e.g., a maximum number of generations is reached).

  return The best *ADS* up to this point.

BP manages the whole genotype at the chromosome level by recognizing it as a single unit that serves the desired goal. Second, the gene level, as in classical GP, considers the tree as the unit where the genetic operation is performed. Therefore, BP allows the creation of new genetic operators inspired by gene and chromosomal biological mutations, as well as the computational crossovers, each one operating at a different level, see Figure 2. These genetic operators are selected according to a probability that is defined following the scheme proposed by Koza, where each operation is computed independently, but their addition of probabilities is one. Hence, the probability of crossover at gene and chromosome levels is 0.4, while the mutation probability at both levels is 0.1.

The next step is the selection of one or two ADSs using the roulette-wheel approach. Thus, the best ADS is kept in the following generation, and the genetic recombination is repeated until a whole new population is created. Finally, the evolutionary loop finishes when a total number of generations, N=30, is reached; Table 5 provides the BP parameters for the experiments. Therefore, once the training stage ends and the ADS with the best fitness is obtained; then, the testing stage starts. Hence, the fittest ADS is tested using a different image database, known as the testing database.

5   Experiments and Results

This section describes the experimental results that were obtained after evolving the ADSs. Also, implementation details are discussed about BP for the sake of clarity. Experiments were performed in a Dell Precision T7500 Workstation, Intel Xeon 8 Core, NVIDIA Quadro FX 3800, and Linux OpenSUSE 11.1 operating system. The following experiments are divided in two parts, according to the goal that the ADS is attempting to reach. The BP is basically the same for both experiments, the only parts that change are the fitness function, which encodes the goal, and the set of images utilized for training that represents the problem. The fitness function and database applied through BP correspond to the characterization of the purpose, the answer to the question: What are the individuals for? In other words, it is the way in which the purpose is implemented as a computer program.

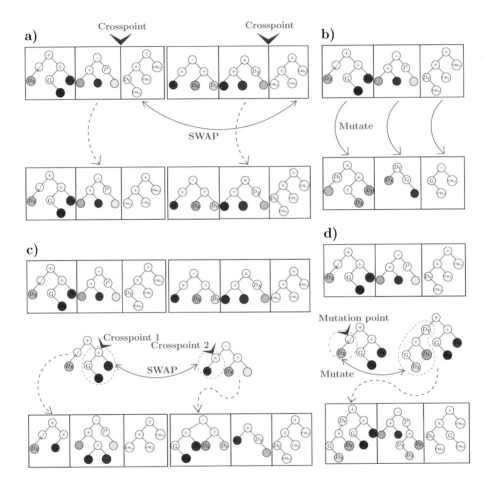

Fig. 2. This figure depicts the genetic operations of crossover and mutation. a) and b) illustrate the crossover and mutation, respectively, at the chromosome level; while, c) and d) show a visual representation of crossover and mutation at the gene level.

Table 5. Initialization values for the BP algorithm

Parameters	Description
Generations	30
Population size	30 individuals
Initialization	Ramped Half-and-Half
Crossover at chromosome level	0.4
Crossover at gene level	0.4
Mutation at chromosome level	0.1
Mutation at gene level	0.1
Tree depth	Dynamic depth selection
Dynamic max depth	7 levels
Real max depth	9 levels
Selection	Roulette-wheel
Elitism	Keep the Best Individual

5.1 Evolution of ADSs for Aiming Scene Novelty

The first set of experiments is designed in terms of *visual search*, which is commonly applied like in classical research devoted to visual attention. In this way, the tests are designed to obtain through artificial evolution an ADS that is

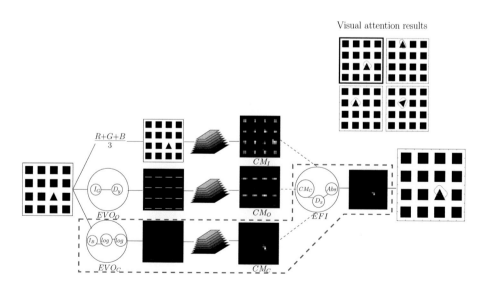

Fig. 3. Bottom-Up image testing of novelty. This figure depicts the best visual attention program that was evolved with brain programming to attend the triangle.

specially adapted to find the novelty, or asymmetries, in a simple set of images of the kind that are used in psychophysical studies.

Search of Appearance Novelty. The first experiment was conceived with the aim of obtaining an ADS capable of centering attention with respect to appearance novelty. Figure 3 shows the $ADS_{triangle}$ that was obtained by the BP strategy. We remark that the $ADS_{triangle}$ utilizes only the color dimension that proposes to regularize the blue band of the image through the logarithm function. This process reduces the contrast between the black and white areas, and as a result, the regions around the triangle are highlighted after the central-surround processing and evolved feature integration steps. Thus, the ADS obtained by BP is listed below:

$$EVO_O = D_y(I_G)$$
$$EVO_C = log(log(I_B))$$
$$EFI = \|D_x(CM_C)\| \quad .$$

(1)

Thus, during evolution only one image per training was used. Note that this image shows the highlighted black square at the top-right corner of Figure 3. The remaining images illustrate the results achieved during a set of preliminary tests considering rotation and translation; indeed, the triangle was correctly focused.

Search of Size Novelty. The experiment described next is noteworthy because, according to the literature, it has not been solved previously by any computational method applied to visual attention. A possible reason may be due to the overlook in the study of the feature size, and consequently the lack of a suitable choice of functions within the problem statement. Thus, in order to develop this experiment we are proposing to increase as an extra dimension the property of shape (EVO_S), which will be computed, as explained in Section 3.1, through the fundamental operations of mathematical morphology. Next, the results obtained along 9 different executions are illustrated below.

The design of this experiment is based on neurological studies of BU attention using a set of images containing an object that is different from the rest. Thereby, the training set consists of 3 images containing an object with bigger size, see Figure 4.

In Figure 5 the statistics presents the BP development along 30 generations considering: a) the fitness, b) the measure of population diversity, and finally, c) and d) to measure the population complexity. Note, that in Figure 5 a), the average fitness of the population is constantly improved, same for the best and median fitness. This means that BP obtains better ADSs while focusing on bigger shapes as evolution progresses. The Figure 5-b) depicts the classical trend towards convergence and loss of diversity in the evolutionary run. BP as any evolutionary algorithm describes the same behavior due to the natural selection principle. Thus, diversity was measured with the Hamming distance among individuals. In this case, between the fitness of two selected ADSs. This

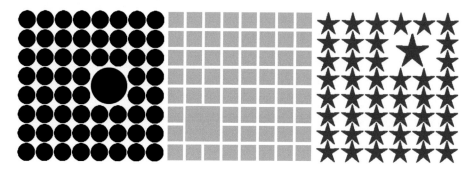

Fig. 4. Training set of images used by BP to obtain an ADS for focusing size novelty

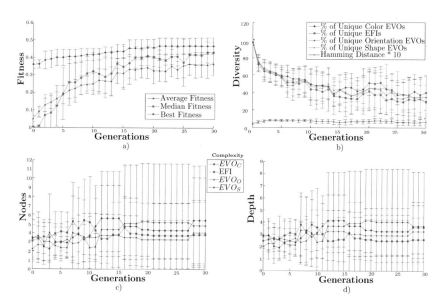

Fig. 5. Brain programming statistics along the execution of 9 experiments for search of size novelty. a) Fitness chart shows average, median, and best fitness, b) Diversity of population shows the percentage of uniqueness of $EVOs$ and EFI operators, as well as the Hamming distance among the individuals of the population; c) and d) depict the structure complexity of the $EVOs$ and EFI operators based on the amount of nodes and depth respectively.

measure show that even when uniqueness decreases the difference among the performance of the ADSs remain almost constant. On the other hand, owing to the flexible representation of tree-based genotype; algorithms solutions may grow too big without any improvement in their performance and generalization-ability. Nevertheless, Figures 5-c) and 5-d) show that for this BP experiment the complexity stays almost the same during evolution while fitness grows.

Afterwards, the best individuals, obtained for each of the BP runs, were tested in a different set of images. The ADS_{Size} with the highest performance during the testing stage is shown in Equation (2). Note, that this ADS_{Size} applies only the feature of orientation, $EFI = CM_O$; while EVOs not needed are shown in red. In this way, the EVO_O consists of the derivatives along y and x direction of the complement image K, followed by a 11.11 root square operation, The resulting information is enough to focus correctly on the bigger objects in most of the testing images. The results in the testing stage are shown in Figure 6, where ADS_{Size} focuses properly the object of interest in 8 out of 9 images.

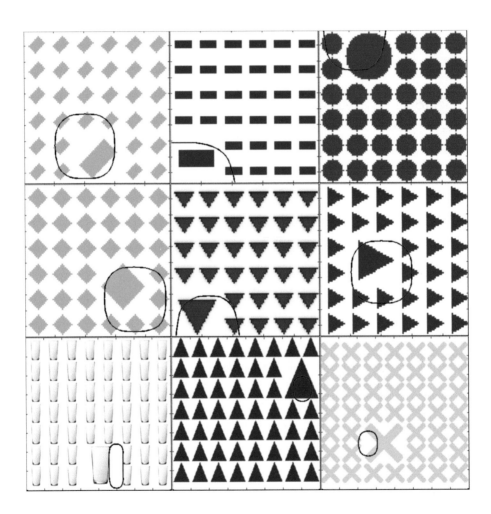

Fig. 6. Results of the best ADS_{Size} along the 9 executions during the testing stage

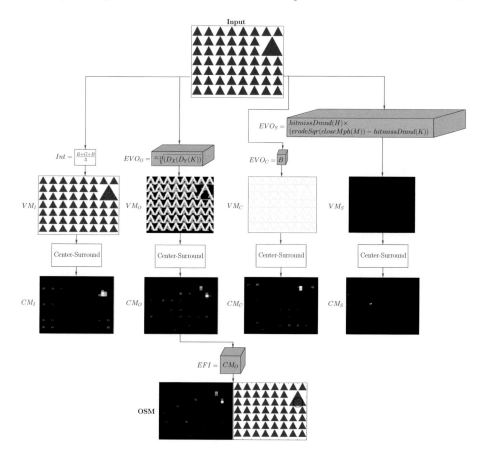

Fig. 7. Example about the functionality of the best ADS$_{Size}$ applied over a testing image

$$EVO_O = \sqrt[11.11]{(D_X(D_Y(K)))}$$
$$EVO_C = B$$
$$EVO_S = hitmissDmnd(H) \times (erodeSqr(closeMph(M)) - hitmissDmnd(K))$$
$$EFI = CM_O$$

$$(2)$$

An example of how the ADS$_{Size}$ behaves during the computation of the saliency region is shown in the Figure 7. Although, all operations were computed, this ADS$_{Size}$ utilizes only the CM_O. Therefore, the whole diagram can be reduced with a significant amount of computational saving. In fact, the best way of executing the whole program is to read the EFI in the first place.

Search for Missing Novelty. Again, the search for missing novelty, as well as size novelty, has not been solved previously by any computational method devoted to the solution of visual attention. In this subsection the results obtained for 10 different runs are described next.

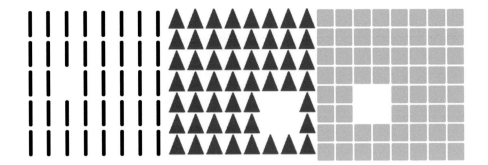

Fig. 8. These figures show the set of images used during training of BP considering the missing novelty problem

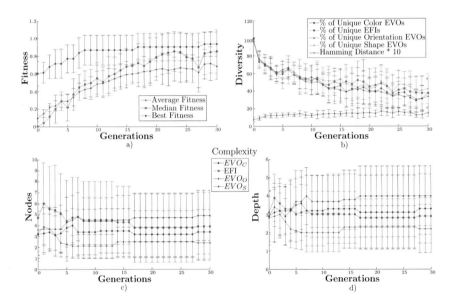

Fig. 9. Brain programming statistics along 10 executions considering the search of empty spaces. a) Fitness chart showing average, median, and best fitness population; b) diversity using the percentage of uniqueness for *EVOs* and *EFI* operators, as well as the Hamming distance among the individuals in the population; c) and d) depict the complexity of the structure for the EVOs and *EFI* operators based on the number of nodes and depth respectively.

The design of this experiment is based on the neurological studies of BU attention using a set of images containing a white space, which is normally interpreted as a missing object. In particular, the set of training images is made of 3 images containing a white region, see Figure 8.

Fig. 10. Results of the best ADS$_{Miss}$ obtained along 10 executions of the testing stage

Figure 9 depicts: a) the average, best, and median fitness of the population, whose results means that BP efficiently optimize ADSs for focusing blank spaces, the graph b) depicts a decreasing trend in the uniqueness of the different operators, similarly, diversity was measured by Hamming distance among individuals. In this case, between the fitness of the ADSs whose measure shows that even when the uniqueness decreases the difference level among the performance of

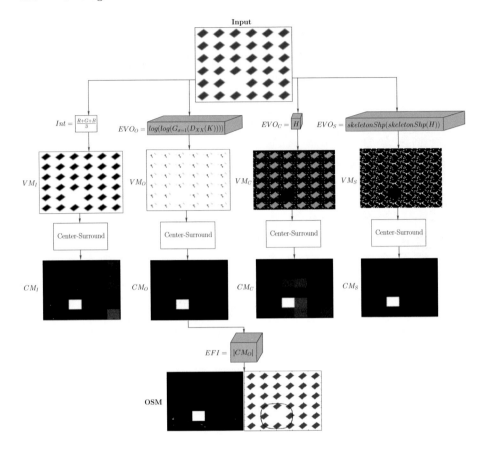

Fig. 11. Example of the functionality of the best ADS$_{Miss}$ applied over a testing image

ADSs exhibits a slight increment. Finally, c) and d) show that for this BP experiment the complexity remains almost constant during evolution while fitness population increases.

Afterwards, the best individuals obtained from each of the BP executions are tested on a different set of images. The ADS$_{Miss}$ exhibiting the highest performance during the testing stage is shown in Equation (3). Note, that this ADS$_{Miss}$ applies only the feature of orientation, note the EFI, while the EVOs that are not used are shown in red. The EVO$_O$ consists of a double derivative along the direction X followed by a Gaussian filter and two logarithm operations. Finally, in the feature integration stage an absolute function was applied to the conspicuity map of orientation CM_O. The results in the testing stage are shown in Figure 10, where the ADS$_{Miss}$ focused properly on the missed object in 15 out of 17 images.

Fig. 12. Training set of images used by BP to obtain an ADS for focusing the red can target

$$EVO_O = log(log(G_{\sigma=1}(D_{XX}(K))))$$
$$EVO_C = H$$
$$EVO_S = skeletonShp(skeletonShp(H)) \tag{3}$$
$$EFI = |CM_O|$$

Figure 11 depicts the information for rightly computing the region of the missing object. Although, in this example the whole ADS_{Miss} is shown only for illustrative purposes, since it only uses the CM_O.

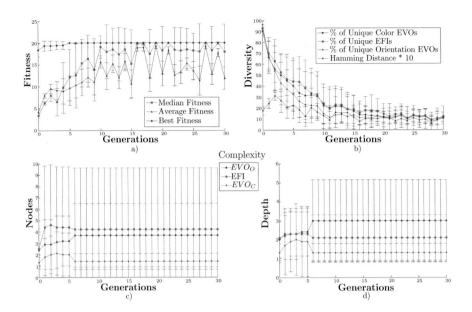

Fig. 13. Brain programming statistics along the execution of 10 experiments for the attention of red can. a) Fitness chart shows average, median, and best fitness, b) diversity of population shows the percentage of uniqueness of $EVOs$ and EFI operators as well as the Hamming distance among the individuals of population, c) and d) depict the complexity of the structure of the $EVOs$ and EFI operators based on the number of nodes and depth respectively.

5.2   Evolution of ADSs for Aiming Specific Targets

In this section, the final ADS_{Can} and their performance are presented for the case of the TD tasks. Figure 13 provides the statistics of the TD runs: the chart 13-a) shows that the average, best, and median fitness of the population improved quickly, considering that the 10 experiments score an ADS with a maximum fitness in the 5th generation or earlier. This means that BP find easily the optimal ADS for focusing the red can. Figure 13-b) demonstrates a decreasing tendency regarding the diversity of multiple operators scoring lower levels with around 15% of uniqueness. This phenomenon is possibly obtained due to an early convergence of the BP process. Consequently, the diversity measured by the Hamming distance scores a decreasing tendency along the whole run. As a result, we can observe how the complexity remain constant after the fifth generation, see Figures 13-c) and 13-d).

Finally, we would like to show some experiments to illustrate that for the red can target a solution could be attained without changing the proposed computational framework. During the training stage, the ADS_{Can} is able to detect the object of interest, in this case the red can, with a successful rate of 100% considering 44 images, see Figure 12. Moreover, during the testing stage the ADS_{Coke}

Fig. 14. Results of the best ADS$_{Can}$ along 10 executions of the testing stage

is able to detect the object of interest with a rate of 91.52% using 59 images. Hence, from 59 test images the coke was detected in 54 occasions. Moreover, the percentage of detection increases after considering a second attempt since the red can was correctly detected in 4 additional images; scoring a total of 58 images that represent the 98.3% of the total, see Figure 14.

Next, the ADS$_{Can}$ obtained by BP is specified in Equation (4). This expression is complex and therefore, it is difficult to decompose its functionality.

$$
\begin{aligned}
EVO_O &= G_{\sigma=1}(M) \\
EVO_C &= (Exp(R) \times \frac{M}{Complement(H)}) \times Exp(H) \\
EFI &= (((D_{YY}(CM_C) + D_{XX}(CM_O)) - Exp(D_{XXX}(CM_I))) \\
&\quad - G_{\sigma=1}(G_{\sigma=2}(G_{\sigma=1}(\sqrt{\frac{G_{\sigma=1}(G_{\sigma=1}(CM_O))}{G_{\sigma=1}(CM_C)}}))))
\end{aligned}
\tag{4}
$$

The obtained ADS$_{Can}$ is complex and has a very good performance. In this way, the Figure 15 is useful to analyze how it works; with such example it is

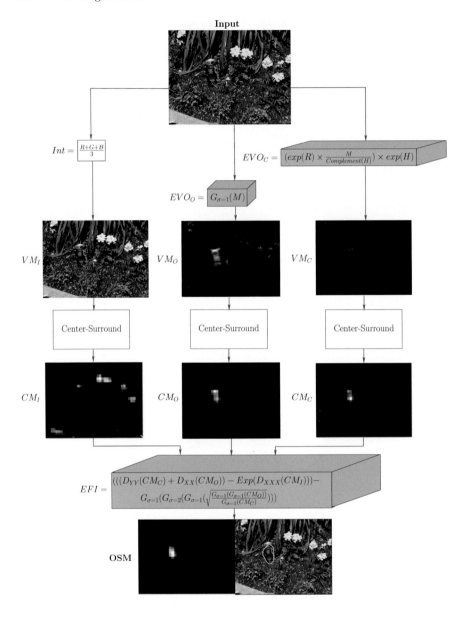

Fig. 15. Example of the functionality of the best ADS$_{Can}$ applied over a testing image

possible to see that the CM_O and CM_C highlight the red can since both are added in order to enhance even more the object of interest. Moreover, the CM_I enhance information that is uninteresting and that is subtracted during the feature integration stage and producing a decrease in the level of noise.

6   Conclusions

This work presents a new and useful approach for understanding visual attention. The experiments are motivated by new ideas about purposive evolution and brain programming. The results confirm that the BP is a powerful methodology that is capable of obtaining ADSs that can be seen as "top-down"or "bottom-up"computational models of the visual attention system that are capable of solving the visual attention problem. Moreover, original programs that solve the size and missing pop-out task were obtained by our approach, and to our knowledge it is the first time to be achieved. Also, the incorporation of shape dimension, carried out with morphological operations, is an original contribution to the research in visual attention. As a conclusion, for some tasks it is not necessary to compute all features; thus, simplifying the ADSs final structure.

Acknowledgements. This research was founded by CONACyT through the Project 155045 - "Evolución de Cerebros Artificiales en Visión por Computadora". Second author supported by scholarship 164678 from CONACyT. This research was also supported by TESE through the project DIMI-MCIM-004/08.

References

1. Atmosukarto, I., Shapiro, L.G., Heike, C.: The use of genetic programming for learning 3d craniofacial shape quantifications. In: Proceedings of the 2010 20th International Conference on Pattern Recognition, ICPR 2010, pp. 2444–2447 (2010)
2. Ayala, F.J.: Teleological explanations in evolutionary biology. Philosophy of Science 37(1), 1–15 (1970)
3. Baluch, F., Itti, L.: Mechanisms of top-down attention. Trends in Neurosciences 34(4), 210–224 (2011)
4. Barton, R.A.: Visual specialization and brain evolution in primates. Proceedings of the Royal Society of London Series B-Biological Sciences 265(1409), 1933–1937 (1998)
5. Borji, A., Sihite, D., Itti, L.: Computational modeling of top-down visual attention in interactive environments. In: Proceedings of the British Machine Vision Conference, pp. 85.1–85.12. BMVA Press (2011)
6. Corbetta, M., Shulman, G.L.: Control of goal-directed and stimulus-driven attention in the brain. Nature Reviews Neuroscience 3(3), 201–215 (2002)
7. Desimone, R., Duncan, J.: Neural mechanisms of selective visual attention. Annual Reviews 18, 193–222 (1995)
8. Feldman, J.A.: Four frames suffice: A provisional model of vision and space. Behavioral and Brain Sciences 8(02), 265–289 (1985)
9. Frintrop, S.: VOCUS: A Visual Attention System for Object Detection and Goal-Directed Search. LNCS (LNAI), vol. 3899. Springer, Heidelberg (2006)
10. Gottlieb, J.: From thought to action: the parietal cortex as a bridge between perception, action, and cognition. Neuron 53(1), 9–16 (2007)
11. Hubel, D.H., Wiesel, T.N.: Receptive fields of single neurons in the cat's striate cortex. Journal of Physiology 148, 574–591 (1959)
12. Itti, L., Koch, C., Niebur, E.: A model of saliency-based visual attention for rapid scene analysis. IEEE Trans. Pattern Anal. Mach. Intell. 20(11), 1254–1259 (1998)

13. Kant, I.: Critique of the power of judgment. Cambridge University Press (2000)
14. Koch, C., Ullman, S.: Shifts in selective visual attention: towards the underlying neural circuitry. Human Neurobiology 4(4), 219–227 (1985)
15. Lennox, J.G.: Darwing was a teleologist. Biology and Philosophy 8, 409–421 (1993)
16. Milanese, R.: Detecting salient regions in an image: from biological evidence to computer implementation. PhD thesis, Department of Computer Science, University of Genova, Switzerland (December 1993)
17. Milner, D., Goodale, M.A.: The visual brain in action, 2nd edn. Oxford University Press (December 1995)
18. Olague, G.: Evolutionary Computer Vision. In: The First Footprints. Springer (to appear)
19. Oliva, A., Torralba, A., Castelhano, M., Henderson, J.: Top-down control of visual attention in object detection. In: International Conference on Image Processing, vol. 1, pp. 253–256 (September 2003)
20. Pérez, C.B., Olague, G.: Learning invariant region descriptor operators with genetic programming and the F-measure. In: Pattern Recognition ICPR 19th International Conference, pp. 1–4 (2008)
21. Peters, R.J., Itti, L.: Beyond bottom-up: Incorporating task-dependent influences into a computational model of spatial attention. In: Proc. IEEE Conference on Computer Vision and Pattern Recognition (CVPR), Minneapolis, MN (June 2007)
22. Posner, M.I., Snyder, C.R., Davidson, B.J.: Attention and the detection of signals. Journal of Experimental Psychology 109(2), 160–174 (1980)
23. Rensink, R.A.: The dynamic representation of scenes. Visual Cognition (2000)
24. Rensink, R.A.: Seeing, sensing and scrutinizing. Vision Research 40, 1469–1487 (2000)
25. Trehub, A.: The Cognitive Brain. Introduces. MIT Press, Cambridge (1991)
26. Treisman, A.M., Gelade, G.: A feature-integration theory of attention. Cognitive Psychology 12(1), 97–136 (1980)
27. Ungerleider, L.G., Mishkin, M.: Two cortical visual systems. In: Ingle, D.J., Goodale, M.A. (eds.) Analysis of Visual Behavior, pp. 549–585. MIT Press, Cambridge (1982)
28. Walther, D., Koch, C.: Modeling attention to salient proto-objects. Neural Networks 19(9), 1395–1407 (2006)

Part III
Multi-objective Optimization

Time Complexity and Zeros of the Hypervolume Indicator Gradient Field

Michael Emmerich and André Deutz

Leiden University, Leiden Institute for Advanced Computer Science
2333 CA Leiden, The Netherlands
{emmerich,deutz}@liacs.nl
http://natcomp.liacs.nl

Abstract. In multi-objective optimization the hypervolume indicator is a measure for the size of the space within a reference set that is dominated by a set of μ points. It is a common performance indicator for judging the quality of Pareto front approximations. As it does not require a-priori knowledge of the Pareto front it can also be used in a straightforward manner for guiding the search for finite approximations to the Pareto front in multi-objective optimization algorithm design.

In this paper we discuss properties of the gradient of the hypervolume indicator at vectors that represent approximation sets to the Pareto front. An expression for relating this gradient to the objective function values at the solutions in the approximation set and their partial derivatives is described for arbitrary dimensions $m \geq 2$ as well as an algorithm to compute the gradient field efficiently based on this information. We show that in the bi-objective and tri-objective case these algorithms are asymptotically optimal with time complexity in $\Theta(\mu d + \mu \log \mu)$ for d being the dimension of the search space and μ being the number of points in the approximation set. For the case of four objective functions the time complexity is shown to be in $\mathcal{O}(\mu d + \mu^2)$. The tight computation schemes reveal fundamental structural properties of this gradient field that can be used to identify zeros of the gradient field. This paves the way for the formulation of stopping conditions and candidates for optimal approximation sets in multi-objective optimization.

Keywords: Set Oriented Optimization, Multiobjective Gradient, Hypervolume Indicator, Computational Complexity, Optimality Conditions.

1 Introduction

The gradient field assigns to each vector in the search space (or decision space) a vector of all partial derivatives at this vector that is called the gradient at this point. Gradients play an important role in the formulation of optimization algorithms, as they are vectors that point in the direction where function values will increase the most and thus can guide the search towards better solutions. Moreover, for differentiable functions the gradient at local optima is zero, which can be used to identify candidates for local optima.

O. Schütze et al. (eds.), *EVOLVE - A Bridge between Probability, Set Oriented Numerics, and Evolutionary Computation III,* Studies in Computational Intelligence 500, DOI: 10.1007/978-3-319-01460-9_8, © Springer International Publishing Switzerland 2014

The problem of solving multi-objective optimization problems, is often restated as finding a finite approximation set to the Pareto front of the problem. In this case the hypervolume indicator provides a figure of merit for an approximation set. Loosely speaking, it measures the volume of the subspace that is Pareto dominated by the approximation set. The hypervolume indicator gradient at a set of decision vectors points in the direction that locally yields maximal improvement of this indicator by simultaneously updating all points. It was first described in [1], but analysis and computation schemes were mainly restricted to the bi-objective case. This chapter presents a substantially extended analysis and efficient algorithms for computing the hypervolume indicator gradient field. In the bi- and tri-objective cases these algorithms are even asymptotically optimal. In particular the following research questions will be addressed:

Given information on the objective function vectors and partial derivatives of the objective functions for all points in the approximation set:

- Can we concisely define the hypervolume indicator gradient field and the points where it is defined for an arbitrary number of objective functions?
- Can structural properties of the gradient expression be exploited to find efficient algorithms for computing the hypervolume gradient?
- Can these structural properties be used to identify compact equations for the zeros of the hypervolume gradient field?

As will be shown, the answer to all three questions is affirmative.

In the following discussion we will first establish a formal framework for defining the hypervolume indicator gradient at an approximation set. Actually, we will be talking about two gradient fields:

1. The gradient field for the mapping from a set of decision vectors to the hypervolume indicator
2. The gradient field for the mapping from a set of objective vectors to the hypervolume indicator

We will proceed with the definition of these gradient fields and identify at which domains consisting of approximation sets the gradient fields are well-defined. Efficient algorithms for the computation of the gradient field at an approximation set will be provided, for both mappings. Their asymptotic optimality for the bi- and tri-objective case will be proven. Finally, a locality property of the hypervolume indicator will be discussed. It yields concise formulations of conditions of points where the gradient field of the first mapping obtains values of zero. This can be used in optimality conditions. The same property gives rise to a new interpretation of the hypervolume indicator gradient field and a technique for its visualization. The final section is also devoted to the discussion of implications of the new theoretical results for set-oriented multi-objective optimization in the future.

2   Related Work

The idea to use gradient information in multi-objective optimization is not new.

Fliege [2] suggests a steepest descent method that searches within the cone of dominating solutions in the direction where the net decrease of objective function values is expected to be maximal among all vectors with a given length, added to the current variable vector. This direction, obtained by quadratic maximization based on the Jacobian (the matrix of the objective functions gradients), is termed *multi-criterion gradient*. Variations and generalizations of this approach have been proposed by Brown and Smith [3] and Bosman and de Jong [4]. More recently a gradient based method that approximates the gradient from points that are generated in an evolutionary search in [5] was suggested. A similar line of research is given by methods that generate non-dominated points by linear combinations of the negative gradients with positive weights [6,7]. For small step-sizes this yields non-dominated or dominating solutions. Thereby, the Euler method is used to integrate along a path of such solutions. Recently, these methods have been hybridized for evolutionary multi-criterion optimization by Shukla et al. [8] by computing favorable directions for generating offspring individuals. Unlike the aforementioned methods, homotopy and continuation methods as described by Hillermeier [9] and Schütze et al. [10] use gradient-based search not in the first place to move search points closer to the Pareto front, but for finding a well-distributed set of points covering the Pareto fronts. The basic idea is to gradually extend the manifold around a given Karush-Kuhn-Tucker point. This way, given a smooth and connected Pareto front, accurate approximations can be achieved. A technique called *directed search* uses gradient information to steer the search in a desired direction given by a vector in the objective space [11].

In this chapter we will further explore an alternative use of gradients in multi-objective optimization that was proposed in [1]. Here the gradient field is formulated on the (μd)-dimensional space of concatenated sets of μ decision vectors in $\mathbb{R}^d$ or, respectively, at multi-sets of decision vectors $\mathbb{R}^{\mu d}$. Following this paradigm, *the improvement of a single decision vector is measured explicitly and solely in how much this vector improves with respect to its contribution to a scalar performance measure stated on an entire set of decision vectors.*

3   Formal Definition of the Hypervolume Indicator Gradient Field

A central concept in this work is that of a gradient at a vector and that of a gradient field. To avoid ambiguity of language we will provide elementary definitions, first.

3.1   Gradient at a Vector and Gradient Field

We introduce partial derivatives via one-sided partial derivatives for a function $\varphi : \mathbb{R}^n \to \mathbb{R}$.

$$\frac{\partial_+ \varphi}{\partial x_i}(\mathbf{x}) = \lim_{t \downarrow 0} \frac{\varphi(\mathbf{x} + t\mathbf{e}_i) - \varphi(\mathbf{x})}{t}$$

denotes the right one-sided partial derivative at $\mathbf{x}$ for x_i, and

$$\frac{\partial_- \varphi}{\partial x_i}(\mathbf{x}) = \lim_{t \uparrow 0} \frac{\varphi(\mathbf{x} + t\mathbf{e}_i) - \varphi(\mathbf{x})}{t}$$

is the left one-sided partial derivative. If both values exists at a point $\mathbf{x}$ and are equal, we denote by

$$\frac{\partial\varphi}{\partial x_i}(\mathbf{x}) := \frac{\partial_+\varphi}{\partial x_i}(\mathbf{x}) = \frac{\partial_-\varphi}{\partial x_i}(\mathbf{x})$$

the partial derivative at $\mathbf{x}$ with respect to x_i.

The gradient of a function $\mathbb{R}^n \to \mathbb{R}$ at a vector is a vector pointing in the direction of the steepest ascent at that point. The steepness of the slope is given by the length of this vector. It is defined via partial derivatives as:

$$\nabla\varphi(\mathbf{x}) := \left(\frac{\partial\varphi}{\partial x_1}(\mathbf{x}), \ldots, \frac{\partial\varphi}{\partial x_n}(\mathbf{x}) \right)^\top. \tag{1}$$

The function $\nabla\varphi : \mathbb{R}^n \to \mathbb{R}^n$ is commonly referred to as the *gradient field* associated to φ.

3.2   Multi-objective Optimization, Efficient Set, and Pareto Front

In multi-objective (or: multicriteria) optimization, we consider an m-tuple of functions

$$(f_1 : \mathbb{R}^d \to \mathbb{R}, \ldots\ldots, f_k : \mathbb{R}^d \to \mathbb{R}, \ldots\ldots, f_m : \mathbb{R}^d \to \mathbb{R}),$$

each function of which is to be minimized or maximized. Without loss of generality, in the following we assume the goal is maximization. We denote by $\mathbf{f} : \mathbb{R}^d \to \mathbb{R}^m$ the corresponding vector valued function $(f_1, \ldots, f_m)^\top$. The practically very important special cases $m = 2$ and $m = 3$ are called bi-objective (or: bicriteria) and tri-objective (or: tricriteria) problems.

In the following discussion it will be important to clearly distinguish between decision vectors $\mathbf{x} \in \mathbb{R}^d$, that is the domain of $\mathbf{f}$ or *decision space*, and objective vectors in $\mathbf{y} \in \mathbb{R}^m$, that is the co-domain of $\mathbf{f}$ or *objective space*. As $\mathbf{f}$ is not necessarily surjective the following definition is made: An objective vector $\mathbf{y}$ is *attainable* if $\mathbf{y} = \mathbf{f}(\mathbf{x})$ for some $\mathbf{x} \in \mathbb{R}^d$. The set of attainable objective vectors is termed *attainable objective space*.

The above problem of multi-objective optimization is not well stated, as it is not clear how to deal with *conflicting objective functions*, that is pairs f_k and $f_{k'}$ with $\arg\min_{\mathbf{x}\in\mathbb{R}^d}(f_k) \cap \arg\min_{\mathbf{x}\in\mathbb{R}^d}(f_{k'}) = \emptyset$. However, Pareto dominance establishes a partial order on the objective space. The maximal elements of this partial order for the attainable objective space we will term *Pareto optimal objective vectors* and their pre-images with respect to $\mathbf{f}$ we will term *efficient decision vectors*. Accordingly, the set of all Pareto optimal objective vectors we term *Pareto front*, whereas the *efficient set* will be the set of all efficient decision vectors. See also Ehrgott [12] for these definitions.

In Pareto optimization we are interested in finding the efficient set and Pareto front for $\mathbf{f}$. The Pareto front is interesting, because it reveals the nature of the trade-off between different objectives and contains all objective vectors that cannot be strictly improved anymore without additional statements about preferences.

Remark 1. Note that we restrict ourselves here to the continuous and uncon-strained case, but definitions can be generalized in a straightforward way to decision spaces with (integrity) constraints. This does however not hold for the gradient computations that will be discussed in this paper.

In continuous multi-objective optimization we face the problem that the effi-cient set and the Pareto front of a function can be innumerably large sets. One approach is to approximate the Pareto front with a finite multi-set of, say μ, attainable objective vectors[1]. We will term a multi-set Y of μ solutions in the attainable objective space an approximation set to the Pareto front , and a multi-set X of μ solutions in the decision space an approximation set to the efficient set.

3.3   Hypervolume Indicator

One approach to state optimality of an approximation set in the decision space is to require for an approximation set of maximal hypervolume indicator H. Roughly speaking, this indicator assigns a better (higher) value to approximation sets to the Pareto front that dominate many objective function vectors than to approximation sets that dominate fewer objective vectors. We define

$$\text{DomSet}(Y) = \{\mathbf{y}' \in \mathbb{R}^m | \ \exists \mathbf{y} \in Y : \mathbf{y} \ \text{Pareto dominates} \ \mathbf{y}'\}$$

As this set has infinite measure, its size cannot serve as an indicator. Instead the hypervolume indicator measures the size of the dominated volume within the reference set $[\mathbf{r}, \infty)$ for a reference vector $\mathbf{r} \in \mathbb{R}^m$. Hence, the definition of the *hypervolume indicator* reads:

$$H(Y, \mathbf{r}) = \lambda(\text{DomSet}(Y) \cap [\mathbf{r}, \infty)),$$

and λ denotes the Lebesgue measure on $\mathbb{R}^m$, that is the area of the dominated set in the reference space is measured in case $m = 2$ and its volume in the case $m = 3$. The choice of a proper reference point is a task that is typically delegated to the user. Ideally it should be dominated by all attainable objective vectors. We write $H(Y)$ instead of $H(Y, \mathbf{r})$ if the definition of $\mathbf{r}$ is clear from the context.

For geometrical considerations the following equivalent definition (for $m > 1$) is sometimes more accessible, but requires $\mathbf{r} \leq \mathbf{y}$, componentwise, for all $\mathbf{y} \in Y$:

$$H(Y, \mathbf{r}) = \lambda(\cup_{\mathbf{y} \in Y}[\mathbf{r}, \mathbf{y}]).$$

The hypervolume indicator (or: S-metric) was first introduced as a *unary per-formance indicator* [13,14] and is nowadays also widely used in bounded-size archiving and to guide the search towards the Pareto front. It is commonly used and analyzed in the context of evolutionary multi-objective optimization [15][16], but has hardly been considered so far in deterministic algorithms for finding the

---

[1] Note that the symbol μ is used, as it is a common symbol for denoting the size of a population in evolutionary multi-objective optimization.

Pareto front (cf. [17]). Recently, the hypervolume indicator received attention in computational geometry as it is a special case of Klee's measure problem and it may serve to establish lower complexity bounds for this problem [18]. In general it is likely that the time complexity of the hypervolume indicator is exponential in dimension m, while fast algorithms with subquadratic time complexity in the number of points in the approximation set μ exists for the practically relevant cases with $m = 2$, $m = 3$ (cf. [19]), and $m = 4$ (cf. [20]).

3.4   Gradients at Approximation Sets

As gradients are defined at vectors and not at multi-sets, a mapping from multi-sets to vectors that represent these will be established next. A multi-set X of μ decision vectors in $\mathbb{R}^d$, that may serve as an approximation to the efficient set, is represented as a concatenation of its elements and called a μd-vector. We say a μd-vector

$$\mathbf{X} := (x_1^{(1)}, \dots, x_d^{(1)}, \dots\dots, x_1^{(i)}, \dots, x_d^{(i)}, \dots\dots, x_1^{(\mu)}, \dots, x_d^{(\mu)})^\top \in \mathbb{R}^{\mu \cdot d}$$

represents the multi-set $\{\mathbf{x}^{(1)}, \dots, \mathbf{x}^{(\mu)}\}$. We name the subsequence with upper index $i \in \{1, \dots, \mu\}$, the i-th *subvector* of the μd vector. Accordingly, we can represent multi-sets in $\mathbb{R}^m$, that may serve as approximations to the Pareto front, as μm-vectors $\mathbf{Y}$. We say

$$\mathbf{Y} := (y_1^{(1)}, \dots, y_m^{(1)}, \dots\dots, y_1^{(i)}, \dots, y_m^{(i)}, \dots\dots, y_1^{(\mu)}, \dots, y_m^{(\mu)})^\top \in \mathbb{R}^{\mu \cdot m}$$

represents the multi-set $\{\mathbf{y}^{(1)}, \dots, \mathbf{y}^{(\mu)}\}$. We name the subsequence with upper index $i \in \{1, \dots, \mu\}$, the i-th *subvector* of the μm vector. As μ (the number of points in the approximation set), d (the number of dimensions of the decision space) and m (the number of objective functions) are constant in the search algorithms that we consider, it turns out to be a convenient notational convention.

The mapping we have just defined is, in general, not injective.

Proposition 1. *Every multi-set of size μ with elements in $\mathbb{R}^d$ (or, respectively, $\mathbb{R}^m$) has at least one and at most $\mu!$ representing μd-vectors (or, respectively μm-vectors). Each μd-vector (or, respectively μm-vector) represents exactly one multi-set in $\mathbb{R}^d$ (or, respectively, $\mathbb{R}^m$).*

Proof. The concatenation of vectors from the multi-set can be done in $\mu!$ different orders. Due to duplicates the number of distinguishable μd vectors might be less than $\mu!$.   □

The proposition takes into account the possibility of duplicates in the multi-set, in which case the number of representations will be less than $\mu!$.

To establish a connection between μd-vectors and μm-vectors, define the mapping $\mathbf{F} : \mathbb{R}^{\mu d} \to \mathbb{R}^{\mu m}$ with

$$\mathbf{X} \mapsto (f_1(\mathbf{x}^{(1)}), \dots, f_m(\mathbf{x}^{(1)}), \dots\dots, f_1(\mathbf{x}^{(\mu)}), \dots, f_m(\mathbf{x}^{(\mu)}))$$

Remark 2. The reformulation of multi-sets to concatenated vectors will not be needed in the long run, as we will show that the gradient can be decomposed into subgradients associated with single points. To say it with Wittgenstein's metaphor, our construction serves as a 'ladder' that after we climbed it can be discarded again.

For technical reasons, first the definition of the hypervolume indicator needs to be slightly adapted to be compatible with the vector representation:

$$\mathcal{H}(\mathbf{Y}) = \lambda \left(\bigcup_{i=1,\ldots,\mu} (-\infty, (y_1^{(i)}, \ldots, y_m^{(i)})^\top] \cap [\mathbf{r}, \infty) \right). \tag{2}$$

Proposition 2. *Let* $\mathbf{Y}$ *denote a* μm-*vector that represents some multi-set* Y *in* $\mathbb{R}^m$. *Then* $\mathcal{H}(\mathbf{Y}) = H(Y)$.
Proof. □

For a given μd-vector $\mathbf{X}$ of μ points we define:

$$\mathcal{H}_\mathbf{F}(\mathbf{X}) := \mathcal{H}(\mathbf{F}(\mathbf{X})). \tag{3}$$

The introduced formal framework is sound, as by optimizing $\mathcal{H}_\mathbf{F}$ over the set of μd-vectors we will obtain multi-sets of maximal hypervolume. For precision, the following lemma is stated:

Lemma 1. *Each multi-set of size* μ *that maximizes* $\mathcal{H}_\mathbf{F}$ *is represented by at least one and at most* $\mu!$ *maxima of* $\mathcal{H}_\mathbf{F}$. *A* μd-*vector that is not maximal with respect to* $\mathcal{H}_\mathbf{F}$ *does not represent a maximal multi-set of size* μ *for* $\mathcal{H}_\mathbf{F}$.

Proof. This follows from Propositions 1 and 2. □

4   The Hypervolume Gradient Field

The gradient field $\nabla \mathcal{H}_\mathbf{F}$ is defined by Equation 1 for the mapping $\mathcal{H}_\mathbf{F}$ at any μd-vector where $\mathcal{H}_\mathbf{F}$ is differentiable, that is for any μd-vector for which all partial derivatives with respect to $\mathcal{H}_\mathbf{F}$ are well defined. Analogously, the gradient field $\nabla \mathcal{H}$ is defined by Equation 1 at any μm-vector where $\mathcal{H}$ is differentiable.

We will first look at how the partial derivatives of the gradient field $\nabla \mathcal{H}$ and $\nabla \mathcal{H}_\mathbf{F}$ can be computed given the information (in the points where the functions are partially differentiable):

$$f_k(\mathbf{x}^{(i)}) \text{ for } i = 1, \ldots, \mu; k = 1, \ldots, m$$

and

$$\frac{\partial f_k}{\partial x_j^{(i)}}(\mathbf{x}^{(i)}) \text{ for } i = 1, \ldots, \mu; j = 1, \ldots, m; k = 1, \ldots, m.$$

Subsequently we will classify regions of differentiability.

4.1   C $\mathcal{H}_{\mathbf{F}}$ at a μd-Vector

Using a different notation the mapping $\mathcal{H}_{\mathbf{F}}$ in Equation 3 can be defined by the following composition of mappings:

$$\mathbb{R}^{\mu \cdot d} \underbrace{\xrightarrow{\;\;\mathbf{F}\;\;}}_{\text{decision space to objective space}} \mathbb{R}^{\mu \cdot m} \underbrace{\xrightarrow{\;\;\mathcal{H}\;\;}}_{\text{objective space to single value}} \mathbb{R}. \qquad (4)$$

According to Equation 1 the hypervolume indicator gradient $\nabla \mathcal{H}_{\mathbf{F}}(\mathbf{X})$ of the composition $\mathcal{H}_{\mathbf{F}} = \mathcal{H} \circ \mathbf{F}$ is defined as:

$$\nabla \mathcal{H}_{\mathbf{F}}(\mathbf{X}) = \left(\frac{\partial \mathcal{H}_{\mathbf{F}}(\mathbf{X})}{\partial x_1^{(1)}}, \dots, \frac{\partial \mathcal{H}_{\mathbf{F}}(\mathbf{X})}{\partial x_d^{(1)}}, \dots, \frac{\partial \mathcal{H}_{\mathbf{F}}(\mathbf{X})}{\partial x_1^{(\mu)}}, \dots, \frac{\partial \mathcal{H}_{\mathbf{F}}(\mathbf{X})}{\partial x_d^{(\mu)}} \right)^{\top} \qquad (5)$$

The chain rule provides us with the gradient of $\mathcal{H}_{\mathbf{F}}$ at a point $\mathbf{X}$:

$$\nabla \mathcal{H}_{\mathbf{F}}(\mathbf{X}) = \left(\left(\nabla \mathcal{H} \begin{pmatrix} \mathbf{f}(\mathbf{x}^{(1)}) \\ \mathbf{f}(\mathbf{x}^{(2)}) \\ \cdots \\ \mathbf{f}(\mathbf{x}^{(\mu)}) \end{pmatrix} \right)^{\top} \cdot \begin{pmatrix} \mathbf{f}' \text{ at } \mathbf{x}^{(1)} & 0 & 0 \cdots & 0 \\ 0 & \mathbf{f}' \text{ at } \mathbf{x}^{(2)} & 0 \cdots & 0 \\ \vdots & \vdots & \vdots \cdots & \vdots \\ 0 & 0 & 0\;0 & \mathbf{f}' \text{ at } \mathbf{x}^{(\mu)} \end{pmatrix} \right)^{\top} \qquad (6)$$

To visualize the structure of the composition we give a detailed description:

$$\underbrace{\left(\left(\begin{smallmatrix} \frac{\partial \mathcal{H}}{\partial y_1^{(1)}} \\ \vdots \\ \frac{\partial \mathcal{H}}{\partial y_m^{(1)}} \\ \vdots \\ \frac{\partial \mathcal{H}}{\partial y_1^{(\mu)}} \\ \vdots \\ \frac{\partial \mathcal{H}}{\partial y_m^{(\mu)}} \end{smallmatrix} \right) (\mathbf{F}(\mathbf{X})) \right)^{\top}}_{\nabla \mathcal{H}(\mathbf{F}(\mathbf{X}))} \cdot \underbrace{\begin{pmatrix} \frac{\partial f_1(\mathbf{x}^{(1)})}{\partial x_1^{(1)}} & \cdots & \frac{\partial f_1(\mathbf{x}^{(1)})}{\partial x_d^{(1)}} & 0 \cdots 0 & 0 & \cdots & 0 \\ \vdots & \vdots & \vdots & \vdots \vdots & \vdots & \vdots & \vdots \\ \frac{\partial f_m(\mathbf{x}^{(1)})}{\partial x_1^{(1)}} & \cdots & \frac{\partial f_m(\mathbf{x}^{(1)})}{\partial x_d^{(1)}} & 0 \cdots 0 & 0 & \cdots & 0 \\ 0 & \cdots & 0 & \vdots \cdots \vdots & 0 & \cdots & 0 \\ \vdots & \vdots & \vdots & \vdots \vdots & \vdots & \vdots & \vdots \\ 0 & \cdots & 0 & \vdots \cdots \vdots & 0 & \cdots & 0 \\ 0 & \cdots & 0 & 0 \cdots 0 & \frac{\partial f_1(\mathbf{x}^{(\mu)})}{\partial x_1^{(\mu)}} & \cdots & \frac{\partial f_1(\mathbf{x}^{(\mu)})}{\partial x_d^{(\mu)}} \\ \vdots & \vdots & \vdots & \vdots \vdots & \vdots & \vdots & \vdots \\ 0 & \cdots & 0 & 0 \cdots 0 & \frac{\partial f_m(\mathbf{x}^{(\mu)})}{\partial x_1^{(\mu)}} & \cdots & \frac{\partial f_m(\mathbf{x}^{(\mu)})}{\partial x_d^{(\mu)}} \end{pmatrix}}_{\mathbf{F}'(\mathbf{x}^{(1)}, \dots, \mathbf{x}^{(\mu)})} \qquad (7)$$

It is clear that $\mathbf{F}'(\mathbf{x}^{(1)}, \dots, \mathbf{x}^{(\mu)})$ depends solely on the gradient functions ∇f_i at the subvectors $\mathbf{x}^{(1)}, \dots, \mathbf{x}^{(\mu)}$ that correspond with the decision vectors of the original problem. Hence, if these $m \cdot \mu$ local gradients are known, the Jacobian matrix $\mathbf{F}'(\mathbf{X})$ can be computed.

4.2   Gradient of the Mapping $\mathcal{H}$ at a μm-Vector

The computation of the components $\nabla \mathcal{H}((y_1^{(1)}, \dots, y_m^{(1)}, \dots, y_1^{(\mu)}, \dots, y_m^{(\mu)}))$ can be traced back to a geometrical problem as depicted in Figure 1. In two dimensions these components are simply the lengths of the line segments of the 'staircase' (or attainment curve). For details, see [1].

Let us next focus on the general case $m \geq 2$: Let $\mathbf{Y}$ denote a μm-vector that is given by the mapping $\mathbf{F}$ at some μd-vector $\mathbf{X}$ in which case $\mathbf{Y} = \mathbf{F}(\mathbf{X})$.

We first look at the case of non-duplicate coordinates in $(y_k^{(1)}, \ldots, y_k^{(\mu)})$ for each $k = 1, \ldots, m$ and points that do not occur at the boundary of the reference space $[\mathbf{r}, \infty)$.

Definition 1. *Let $\pi_{1,\ldots,\check{k},\ldots,m}(\mathbf{y}) \in \mathbb{R}^{m-1}$ denote the projection of a subvector $\mathbf{y}$ in the μm-vector onto the coordinates $1, \ldots, \check{k}, \ldots, m$, where $\check{k}$ means that k is omitted.*

Theorem 1. *Let $i \in \{1, \ldots, \mu\}$. Let H_{m-1} denote the hypervolume indicator for the $(m-1)$-dimensional objective space with reference space $[\pi_{1,\ldots,\check{k},\ldots,m}(\mathbf{r}), \infty)$. Let $Y_{(i)}^{>k}$ denote the multi-set of projections $\pi_{1,\ldots,\check{k},\ldots,m}(\mathbf{y}^{(i)})$ of subvectors $\mathbf{y}^{(i)}$ of a μm-vector $\mathbf{Y}$ with a higher k-th coordinate than the k-th coordinate of the subvector $\mathbf{y}^{(i)}$.*

$$\frac{\partial \mathcal{H}_m}{\partial y_k^{(i)}}(\mathbf{Y}) = H_{m-1}(Y_{(i)}^{>k} \cup \{\pi_{1,\ldots,\check{k},\ldots,m}(\mathbf{y}^{(i)})\}) - H_{m-1}(Y_{(i)}^{>k}).$$

Proof. The theorem follows from the geometrical insight that for a sufficiently small Δ a small variation of the m-th coordinate in positive direction by the amount of Δ will cause a linear increment of the hypervolume indicator by the size of a slice, given by the face of the attainment surface [21] adjacent to this point in the $(m-1)$-dimensional projection times Δ. See also Figure 1 and Figure 2 for a visualization of the geometrical construction in 2-D and, respectively, 3-D.

Example 1. The construction in Figure 2 shows, here for $i = 2$ and $k = 3$, that one-sided partial derivatives are equal to areas of the visible face A which is adjacent to the i-th subvector. In this example $\partial_- \mathcal{H}/\partial y_3^{(2)}$ is smaller than $\partial_+ \mathcal{H}/\partial y_3^{(2)}$ and hence $\partial \mathcal{H}/\partial y_3^{(2)}$ is not defined. Also for $y_3^{(3)}$ one-sided partial derivatives are unequal, while for all other coordinates the one-sided partial derivates are equal in the positive and negative coordinate direction and thus the partial derivatives are defined.

4.3   Characterization of the Set of Differentiable Points

Partial derivatives of a μm vector are either positive, zero, or undefined. In case they are undefined, still all one-sided derivatives exist, but are unequal for at least one coordinate of the μm vector. Next, we provide criteria based on properties of subvectors that allow in most cases to decided whether or not a μm vector is differentiable.

Let us partition the multi-set of subvectors of a given μm-vector $\mathbf{Y}$ with respect to Pareto dominance, relative to the other subvectors, and relative to the reference space:

1. Partitioning into subsets based on Pareto dominance relative to the other subvectors in $\mathbf{Y}$

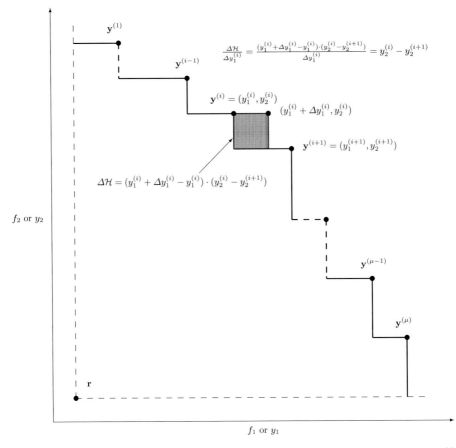

Fig. 1. Geometrical construction used for identifying the partial derivatives $\partial \mathcal{H}/\partial y_j^{(i)}$ of the gradient for $m = 2$ at some non-dominated μm-vector

(a) S: Is the set of strictly dominated subvectors, that is subvectors for which there exists a subvector in $\mathbf{Y}$ that is strictly better in all coordinates.

(b) W: Is the set of weakly dominated subvectors, that is subvectors for which there exists no subvector in $\mathbf{Y}$ that is strictly better in all coordinates and that are Pareto dominated by at least one subvector in $\mathbf{Y}$.

(c) N: Is the set of non-dominated[2] subvectors that in no objective space coordinate have a duplicate value with another non-dominated subvector at this coordinate, e.g. a subvector $(1, 2, 3)^\top$ and another subvector $(3, 2, 1)^\top$ have no duplicate but $(2, 1, 3)^\top$ and $(3, 1, 2)^\top$ have.

(d) D: Is the set of non-dominated subvectors with duplicate coordinates for some objective space coordinate.

---

[2] where non-domination means here Pareto domination with respect to another subvector in $\mathbf{Y}$

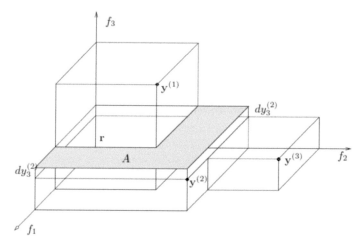

Fig. 2.   Geometrical   construction   for   identifying   a   one-sided   partial derivative $\partial_+ \mathcal{H}/\partial y_3^{(2)}$:= A at   some   non-dominated μd-vector $\mathbf{Y}$ = $(y_1^{(1)}, y_2^{(1)}, y_3^{(1)}, \ldots, y_1^{(3)}, y_2^{(3)}, y_3^{(3)})^T$ for $m = 3$.

2. Partitioning into subsets relative to the reference space $[\mathbf{r}, \infty)$
   (a) I: Is the set of subvectors in the interior of the reference space.
   (b) B: Is the set of subvectors on the boundary of the reference space.
   (c) E: Is the set of subvectors in the exterior of the reference space.

Furthermore, we can partition subvectors with respect to differentiability:

1. Z: Is the set of subvectors for which all partial derivatives are zero.
2. U: Is the set of subvectors for which some partial derivatives are undefined, but as always is the case for the hypervolume indicator $\mathcal{H}$ one-sided partial derivatives exist.
3. P: Is the set of subvectors for which the partial derivatives are all positive.

The relation between these subsets is summarized in the following proposition

Proposition 3

$$Z = E \cup S \tag{8}$$
$$U = D \cup (W \setminus E) \cup (B \setminus S) \tag{9}$$
$$P = N \cap I \tag{10}$$

Proof. In the exterior E and the strictly dominated subspace any differential move of a subvector will leave the hypervolume unchanged, therefore all partial derivatives are zero. In case of $N \cap I$ the size of the face that determines the one-sided partial derivative is the same for the positive and negative direction of a differential move of a single coordinate. It is positive, because the hypervolume will increase (decrease) at the same linear rate when moving the point up or down in the k-th coordinate. It needs to be shown that the rate is positive. The rate is given by the increment of the $m-1$ dimensional hypervolume to the hypervolume of $Y_{(i)}^{>k}$. This increment must be strictly positive, because the projected point is non-dominated with respect to $Y_{(i)}^{>k}$, and when adding a non-dominated point to a set the hypervolume increases (strict monotonicity property [22]). The projected subvector must be non-dominated in the $m-1$ dimensional projection with respect to the subvectors in $Y_{(i)}^{>k}$ because these vectors are already 'better' in the k-th coordinate and points in N must be non-dominated in m dimensions. For U we cannot decide based on the proposition whether all partial derivatives exist, but the one sided derivatives exist as by changing a single coordinate the hypervolume changes at a linear rate (proportional to the size of a $m-1$ dimensional cuboid) or it remains constant. Clearly Z, U and P do not overlap and cover the set of possible subvectors and thus $\{Z, U, P\}$ forms a partition of the set of subvectors. □

Theorem 2. *A μm-vector with partitionings $\{S, W, N, D\}$ and $\{I, B, E\}$ is differentiable, if $U = D \cup (W \setminus E) \cup (B \setminus S) = \emptyset$.*

Proof. Because Z, U, P is a partition, if the condition is satisfied all subvectors are either in Z or in P and therefore their partial derivatives are defined (either zero or positive). □

Remark 3. In three and more dimensions it is possible that all partial derivatives are defined at subvectors that are non-adjacent but have one coordinate in common. An example would be $\mathbf{Y} = ((1, 5, 2) \circ (5, 1, 2) \circ (3, 3, 3))^{\top}$. Here we use $\circ$ as a symbol for *concatenation* of tuples, e.g. $((a, b) \circ (c, d)) = (a, b, c, d)$. An example where partial derivatives are undefined due to duplicate coordinates is given with Figure 2, for the 3-rd subvector and the 2-nd subvector.

Example 2. In Figure 3 a μm-vector with $\mu = 10$ and $m = 2$ is depicted. We obtain these subsets:

Partition based on dominance: $S = \{6, 9, 10\}, W = \{5, 7\}, N = \{1, 2, 3, 4, 8\}$, $D = \emptyset$

Partition based on reference space: $I = \{1, 3, 4, 5, 6, 7\}, B = \{8, 10\}, E = \{2, 9\}$

Partition based on differentiability: $Z = \{2, 6, 9, 10\}, U = \{5, 7, 8\}, P = \{1, 3, 4\}$.

Clearly, only subvectors in $U \neq \emptyset$ might have unequal one-sided partial derivatives. This is indeed the case for the 5-th and 8-th subvector, while for the 7-th subvector the one-sided partial derivatives are equal and zero.

Duplicates among subvectors in the same coordinate can be checked for easily, and whenever they are obtained and the subvectors are neither in S or in E a deeper investigation might be required for checking differentiability. Next, a neccessary condition for differentiability of such μm vectors will be derived for cases where subvectors are in the interior of the reference space $I = (\mathbf{r}, \infty)$.

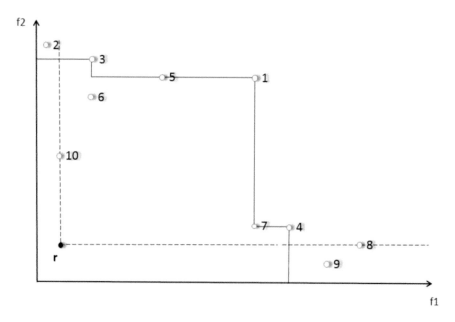

Fig. 3. Differentiability regions

Definition 2. *A subvector* $\mathbf{y}^{(i1)}$ *of* $\mathbf{Y}$ *is said to be* interlaced *with a subvector* $\mathbf{y}^{(i2)}$ *in* $\mathbf{Y}$, *iff* $\exists k \in \{1, ..., m\}$: $y_k^{(i1)} = y_k^{(i2)}$ *and*

$$\lambda(([\mathbf{r}_{\check{k}}, \pi_{1,...,\check{k},...m}(\mathbf{y}^{(i1)})] \cap [\mathbf{r}_{\check{k}}, \pi_{1,...,\check{k},...m}(\mathbf{y}^{(i2)})]) \setminus \mathrm{DomSet}(Y_{(i1)}^{>k})) > \delta \quad (11)$$

for some $\delta > 0$, *where* λ *denotes the* $m-1$ *dimensional Lebesgue measure and* $\mathbf{r}_{\check{k}} = \pi_{1,...,\check{k},...,m}(\mathbf{r})$.

Proposition 4. *If any two subvectors in* I *are interlaced for some* μm-*vector, the function* $\mathcal{H}$ *is not differentiable.*

Proof. If two vectors are interlaced for the k-th coordinate for which the condition is satisfied it clearly holds that $\partial_- \mathcal{H}/y_k^{(i1)} + \delta \leq \partial_+ \mathcal{H}/y_k^{(i1)}$, and hence the two one-sided partial derivatives are not the same. □

It is conjectured that differentiability is given exactly when all subvectors in the interior of the reference space I are mutually non-interlaced. However, a further investigation of this question and the question of how to check the condition in Equation 11 efficiently is left to the future work.

Finally, the following proposition states a sufficient condition for differentiability of $\mathcal{H}_{\mathbf{F}}$.

Proposition 5. *The set of differentiable points of* $\mathcal{H}_{\mathbf{F}}$ *comprises all* μd-*vectors* $\mathbf{X}$ *for which* $\mathcal{H}$ *is differentiable at* $\mathbf{F}(\mathbf{X})$ *and* $\mathbf{f}$ *is differentiable at all subvectors of* $\mathbf{X}$.

Proof. This follows from the well-known fact that the composition of differentiable functions is differentiable and the fact that F is differentiable at $\mathbf{X}$ iff $\mathbf{f}$ is differentiable at each subvector of $\mathbf{X}$.   □

Remark 4. Note, that there are points for which $\mathcal{H}_{\mathbf{F}}$ is differentiable that are not captured in the above proposition. In these cases $\mathbf{f}'$ has at some subvectors of $\mathbf{X}$ zero components. Because of these zero components the one-sidedness of components in $\nabla \mathcal{H}(\mathbf{F}(\mathbf{X}))$ might not influence the differentiability at $\mathbf{X}$, if the position of the zeros matches the position of the one-sided derivatives.

5   Efficient Computation

Next, the computational time complexity of computing the gradient field $\nabla \mathcal{H}_{\mathbf{F}}$ at μd-vectors, given the Jacobian matrices $\mathbf{f}'(\mathbf{x}^{(i)})$, $i = 1, \ldots, \mu$ and $\mathbf{F}(\mathbf{X})$ is discussed. Note that the input data requires memory space in $\Theta(\mu dm)$, and the output data requires memory space in $\Theta(\mu d)$. Only worst case complexities are considered here.

A naïve implementation of the scheme proposed above has super-quadratic complexity in the number of points of the approximation set, because a straightforward computation of Equation 6, that is $\nabla \mathcal{H}(\mathbf{F}(\mathbf{X}))^{\top} \mathbf{F}'(\mathbf{x})$ requires no less than $\mu^2 m^2 d$ multiplications and memory resources proportional to $\mu^2 md$.

The computation of the hypervolume indicator can be done efficiently by utilizing

1. the sparsity of the Jacobian matrix $\mathbf{F}'(\mathbf{X})$, and
2. fast dimension sweep algorithms for incremental hypervolume updates when computing $\nabla \mathcal{H}(\mathbf{Y})$ at a given μm-vector $\mathbf{Y}$.

5.1   Exploiting Sparsity in Matrix Multiplication

An observation from studying the structure in Equation 7 is that many components have a zero value and for each column of the matrix only m components need to be considered in the scalar multiplication with the vector on the right hand side.

Theorem 3. *Given a vector valued objective function* $\mathbf{f} : \mathbb{R}^{\mathbf{d}} \to \mathbb{R}^{\mathbf{m}}$, *a* μd-*vector* $\mathbf{X}$, *the partial derivatives* $\frac{\partial \mathcal{H}}{\partial y_k^i}(\mathbf{F}(\mathbf{X}))$ *and* $\frac{\partial f_k(\mathbf{x}^{(i)})}{\partial x_j^{(j)}}$ *for* $i = 1, ..., \mu$; $j = 1, \ldots, d$; *and* $k = 1, \ldots, m$ *the* μd *components of* $\frac{\partial \mathcal{H}_{\mathbf{F}}}{\partial x_j^{(i)}}(\mathbf{X})$ *can be computed with a computational complexity in* $\mathcal{O}(\mu dm)$ *by means of*

$$\frac{\partial \mathcal{H}_{\mathbf{F}}}{\partial x_j^{(i)}}(\mathbf{X}) = \sum_{k=1}^{m} \frac{\partial \mathcal{H}}{\partial y_k^{(i)}}(\mathbf{F}(\mathbf{X})) \cdot \frac{\partial f_k(\mathbf{x}^{(i)})}{\partial x_j^{(i)}}, i = 1, \ldots, \mu, j = 1, \ldots, d. \quad (12)$$

Proof. This follows immediately when omitting all zero terms in Equation 6.   □

5.2   Dimension Sweep Algorithms for Computing $\nabla \mathcal{H}$

The next goal is to efficiently compute the components of the gradient of the mapping from the objective space to the hypervolume indicator, that is $\frac{\partial \mathcal{H}}{\partial y_k^{(i)}}(\mathbf{Y})$, $i = 1, \ldots, \mu$ and $k = 1, \ldots, m$, for some μm-vector $\mathbf{Y} \in \mathbb{R}^{\mu m}$ after checking for differentiability of $\mathcal{H}$ in $\mathbf{Y} \in \mathbb{R}^{\mu m}$.

Recall, that Theorem 1 states that the partial derivative is given by the incremental change in the dominated hypervolume of the $(m-1)$-dimensional projection, after adding a single point.

Our algorithm is inspired by dimension sweep algorithms for computing the hypervolume indicator as described in [19] and, for 4-D, in [23].

Figure 4 outlines the details of the algorithm to compute hypervolume components. The first part of the algorithms determines all subvectors that evaluate to zero and subvectors with undefined partial derivatives (lines 1-3). This requires a classification of subvectors using Proposition 3. If the set of undefined subvectors is non-empty the μm-vector, $\mathcal{H}$ will be classified as undefined in (cf. Theorem 2). For the Z, U, P partition different sets need to be identified. The non-dominated set can be identified with time complexity in $O(\mu(\log \mu)^{\max(1,m-2)})$ using the algorithm of Kung et al. [24]; all other sets and set-operations can be computed with time complexity in $\mathcal{O}(m\mu \log \mu)$ either using elementary algorithms or based on sorting [25].

In the remainder the algorithm computes gradient components for subvectors in P based on the definition in Theorem 1. This is done by m dimension sweeps, each one computing the partial derivatives of subvectors in P of the k-th objective function.

Following Theorem 1 starting from the subvector with highest k-th coordinate the algorithm adds one by one in descending order of the k-th coordinate the projected subvectors q to a balanced tree data structure T and computes the incremental change in the $(m-1)$-dimensional hypervolume indicator of the set of points processed so far (all points higher in the k-th coordinate as it follows from Theorem 1) caused by this insertion. For the computation it is only required to maintain the set of the non-dominated points in the $(m-1)$-dimensional projection among the points that have been processed so far in the k-th sweep. The tree data structure T is used to maintain this set and quickly identify dominated points to be removed. This way a fast amortized logarithmic-time update schemes (2-D) or amortized linear-time update schemes (3-D) for the hypervolume indicator can be achieved. These update algorithms can be derived from the algorithms described by Beume et al. [19] and, in more than three dimensions, by Guerreiro et al. [23]. For a discussion of the reformulation of these dimension sweep algorithms for computing the m-dimensional indicator, as incremental update schemes for $(m-1)$-dimensional hypervolume indicators, see Hupkens and Emmerich [26]. In the last step of the algorithm's iteration dominated points are removed from the tree. The time for this step amortizes to the cost of identifying a single dominated point, as elements can be removed only once.

Algorithm: GRADMULTISWEEP
Input: μm vector $\mathbf{Y}$ with subvectors $\mathbf{y}^{(1)} \in \mathbb{R}^m, \ldots, \mathbf{y}^{(\mu)} \in \mathbb{R}^m$, reference point $\mathbf{r}$

Output: Partial derivatives $\frac{\partial \mathcal{H}}{\partial y_k^{(i)}}$, $i = 1, \ldots, \mu$; $k = 1, \ldots, m$.

1. Determine the partition Z, U, and P of the subvectors of $\mathbf{Y}$ using Proposition 3.
2. **if** $U \neq \emptyset$ output ("Partial derivatives might be only one-sided in " + U)
3. Assign 0 to all partial derivatives of subvectors in Z.
4. **Remark**: In the remainder compute partial derivatives for all subvectors in P.
5. **For** $k \in \{1, \ldots, m\}$
   (a) Compute P_k as the set of all $(m-1)$-dimensional projections of subvectors of P by omitting their k-th coordinate.
   (b) Add subvectors in P_k in descending order of the k-th coordinate to a queue Q.
   (c) Initialize tree data structure for collecting non-dominated point T as empty.
   (d) **While** Q is not empty:
       i. q ← Lop off first (greatest) element from the queue Q.
       ii. Compute increment $\Delta H(q, \mathtt{T})$ of $(m-1)$-dimensional hypervolume indicator when adding $\mathbf{q}$ to T using efficient update schemes (for $m = 2$ sorting can be used, for $m = 3$ see Beume et al. [19], and for $m \geq 4$ see Guerreiro et al.[23]).
       iii. Set $\frac{\partial \mathcal{H}}{\partial y^{(i(\mathbf{q}))}} = \Delta H(q, T)$, where $i(\mathbf{q})$ is the index that corresponds to the index of the original subvector in $\mathbf{Y}$ of which $\mathbf{q}$ is the projection.
       iv. Add $\mathbf{q}$ to T and remove all elements that are Pareto dominated in the $(m-1)$-dimensional projection by $\mathbf{q}$ from T.

Fig. 4. Computing gradient components

The following theorem summarizes the complexity results of computing gradients of the hypervolume in the objective function space:

Theorem 4. *Given a μm-set $\mathbf{Y}$ of μ concatenated vectors of size m with no duplicate coordinates among subvectors. Then the computation of all components $\partial \mathcal{H}/\partial y_k^{(i)}(\mathbf{Y})$ for $k = 1, \ldots, m$ has a time complexity in $\Theta(\mu \log \mu)$ for $m = 2, 3$ and a time complexity in $\mathcal{O}(\mu^2)$ for $m = 4$.*

Proof. The lower bound of $\Omega(\mu \log \mu)$ for $m = 2$ can be proven by reduction of uniform gap as in [19]. For a given set $\{u_1, \ldots, u_m\}$ we need to represent this set as an instance of to the hypervolume gradient in linear time by duplication of coordinates, yielding $(u_1, -u_1, u_2, -u_2, \ldots, u_\mu, -u_\mu)$. After computing the hypervolume partial derivatives for a reference point $\mathbf{r} = (\min_{i=1,\ldots,\mu}\{u_i\}, -\max_{i=1,\ldots,\mu}\{u_i\})$, the uniform gap is decided positive if and only if all non-zero partial derivatives are the same, which can be checked in a linear number of comparisons.

For proving a lower bound for $m = 3$, we show that there exists a linear time reduction of the hypervolume indicator in two dimensions to the problem of computing the gradient components in three dimensions. As the complexity of computing the hypervolume indicator in two dimensions was proven by Beume et al. [19] to be in $\Omega(\mu \log \mu)$, a time complexity faster than $\Omega(\mu \log \mu)$ would yield a contradiction. The reduction reads as follows: Given μ mutually non-dominated vectors in 2-D, say $\mathbf{u}^{(1)}, \ldots, \mathbf{u}^{(\mu)}$, and assume they are all dominating the 2-D reference point $(r_1, r_2)^\top$. Now we can construct a problem with reference point $(r_1, r_2, 0)^\top$ and a μm-vector $\mathbf{Y} = (u_1^{(1)}, u_2^{(1)}, 1)^\top, \ldots, (u_1^{(1)}, u_2^{(1)}, \mu)^\top$, then $H(\mathbf{u}^{(1)}, \ldots, \mathbf{u}^{(\mu)}) = \sum_{i=1}^{\mu} \frac{\partial \mathcal{H}}{\partial y_3^{(k)}}$.   □

Example 3. This example illustrates the computation of $\nabla \mathcal{H}$ at a μm-vector $\mathbf{Y} = ((12, 11, 7) \circ (1, 3, 7) \circ (3, 10, 8) \circ (14, 4, 5) \circ (6, 12, 4) \circ (-1, 2, 9))^\top$ using Algorithm 4. Reference point is $(0, 0, 0)^\top$. The algorithm first partitions the multiset of subvectors into $U = \emptyset$, $Z = \{(1, 3, 7)^\top, (-1, 2, 9)^\top\}$ and $P = \{(12, 11, 7)^\top$ $(3, 10, 8)^\top$, $(14, 4, 5)^\top$ $(6, 12, 4)^\top\}$. All partial derivatives of the 2nd and 6th subvector are set to zero. Figure 5 visualizes a sweep of P for the final outer loop with index $k = 3$: We initialize the queue as $\mathbf{Q} = [(6, 12)^\top \rightsquigarrow (14, 4)^\top \rightsquigarrow (12, 11)^\top \rightsquigarrow (3, 10)^\top]$. The pictures from the left to the right Figure 5 show the situations right after each iteration of the inner loop. First the algorithm lops off $\mathbf{q}$ at the front of the queue and inserting it to $\mathbf{T}$. The hypervolume update in the $(m-1)$-dimensional projection to f_1 and f_2 is now 30 and the partial derivative $\partial \mathcal{H}/\partial y_3^{(3)}$ is set to this value, because 3 is the upper index of the subvector from which the current $\mathbf{q}$ originated. Now, $\mathbf{Q} = [(6, 12)^\top \rightsquigarrow (14, 4)^\top \rightsquigarrow (12, 11)^\top]$ and the tree contains element $(3, 10)^\top$. In the next iteration $\mathbf{q} = (12, 11)^\top$ is drawn from the queue. The hypervolume update is now 102 and assigned to $\partial \mathcal{H}/\partial y_3^{(1)}$, as 1 is the index of the subvector in $\mathbf{Y}$ from which $\mathbf{q}$ originated. The vector $(3, 10)^\top$ is removed from the tree, because $\mathbf{q} = (12, 11)^\top$ dominates it in the first two dimensions. The next two iterations will not remove points from the tree and the partial derivatives $\partial \mathcal{H}/\partial y_3^{(4)} = 8$ and $\partial \mathcal{H}/\partial y_3^{(5)} = 6$ will be computed in this order. Thereafter the queue is empty and the algorithm terminates.

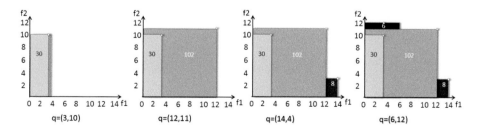

Fig. 5. Gradient computation for a 3-D gradient $\nabla \mathcal{H}(\mathbf{Y})$ for $\mathbf{Y} = ((3, 10, 8) \circ (12, 11, 7) \circ (14, 4, 5) \circ (6, 12, 4))^\top$ and reference point $\mathbf{r} = (0, 0, 0)^\top$

5.3   Time Complexity of Computing the Gradient at a μd-Vector

When putting the results of Theorems 4 and 3 together we obtain that the time complexity in the number of points in the approximation set, μ, is governed by the bounds given in Theorem 4. However, when dealing with a large number of dimensions the influence of m and d might be considerable. The cost for the matrix multiplication is influenced by the search space dimension and scales with $\mathcal{O}(\mu d m)$. Here μd is the same complexity as computing the gradients of all points in the approximation set and thus is at its lower bound.

Theorem 5. *Given an objective function $\mathbf{f}$, a μd vector $\mathbf{X}$, the partial derivatives $\frac{\partial \mathcal{H}}{\partial y_k^i}(\mathbf{F}(\mathbf{X}))$ and $\frac{\partial f_k(\mathbf{x}^{(i)})}{\partial x_i^{(j)}}$ for $i = 1, ..., \mu$; $j = 1, ..., d$; and $k = 1, ..., m$ the time complexity of computing all μd components of $\frac{\partial \mathcal{H}_{\mathbf{F}}}{\partial x_j^{(i)}}(\mathbf{X})$ of the hypervolume gradient $\mathcal{H}_{\mathbf{F}}(\mathbf{X})$ is given by $\Theta(d\mu + \mu \log \mu)$ in $m = 2$ and $m = 3$ dimensions, and by $\mathcal{O}(\mu d + \mu^2)$ in $m = 4$ dimensions.*

Proof. The output size is μd, therefore this is a lower bound for the complexity. Then the result follows from Theorem 3 and Theorem 4 and the fact that m is assumed to be constant. □

6   Gradient Components and Hypervolume Contributions

Revealing the relation between hypervolume contributions of points and the gradient components provides an important insight into the structure of the gradient field, that can yield (1) an alternative algorithm for computing hypervolume contributions, and (2) a concise formulation of an optimality criterion.

The hypervolume contribution $\Delta H(\mathbf{y}, Y)$ of a multi-set Y and a point $y \in Y$ is defined as:

$$\Delta H(\mathbf{y}, Y) = H(Y) - H(Y \setminus \{y\}) \tag{13}$$

Accordingly, define the hypervolume contribution $\Delta \mathcal{H}(i, Y), i = 1, \ldots, \mu$ of the i-th subvector in the μm-vector $\mathbf{Y}$ as the size of the truncated dominated subspace that is dominated by the i-th subvector but not by any other subvector. Putting this into more concrete terms, let $\pi_{1,\ldots,\tilde{i},\ldots,\mu}(\mathbf{Y})$ denote the projection of $\mathbf{y}$ with the i-th subvector removed. Then

$$\Delta \mathcal{H}(i, \mathbf{Y}) = \mathcal{H}(\mathbf{Y}) - \mathcal{H}(\pi_{1,\ldots,\tilde{i},\ldots\mu}(\mathbf{Y}))$$

From the geometrical situation described in Theorem 1 we obtain:

$$\nabla \Delta \mathcal{H}(i, \mathbf{Y}) = \left(\frac{\partial \Delta \mathcal{H}(i, \mathbf{Y})}{\partial y_1^{(i)}}, \ldots, \frac{\partial \Delta \mathcal{H}(i, \mathbf{Y})}{\partial y_m^{(i)}} \right)^{\top} = \tag{14}$$

$$= \left(\frac{\partial \mathcal{H}(\mathbf{Y})}{\partial y_1^{(i)}}, \ldots, \frac{\partial \mathcal{H}(\mathbf{Y})}{\partial y_m^{(i)}} \right)^{\top} \tag{15}$$

Furthermore, let us define the following subgradient at $\mathbf{X}$:

$$\nabla \mathcal{H}_{\mathbf{F}}(i, \mathbf{X}) = \left(\frac{\partial \mathcal{H}_{\mathbf{F}}}{\partial x_1^{(i)}}, \dots, \frac{\partial \mathcal{H}_{\mathbf{F}}}{\partial x_d^{(i)}} \right)^{\top},$$

that is $\mathcal{H}_{\mathbf{F}}(i, \mathbf{X})$ is equal to the i-th subvector of $\mathcal{H}_{\mathbf{F}}(\mathbf{X})$.

Let us recall the equation from Theorem 3:

$$\frac{\partial \mathcal{H}_{\mathbf{F}}}{\partial x_j^{(i)}}(\mathbf{X}) = \sum_{k=1}^{m} \frac{\partial \mathcal{H}}{\partial y_k^{(i)}}(\mathbf{F}(\mathbf{X})) \cdot \frac{\partial f_k(\mathbf{x}^{(i)})}{\partial x_j^{(i)}}, i = 1, \dots, \mu, j = 1, \dots, d. \qquad (16)$$

It can be written in a compact form:

Theorem 6. *Let $\mathbf{f}'(\mathbf{x}^{(i)})$ denote the Jacobian matrix of $\mathbf{f} : \mathbb{R}^d \to \mathbb{R}^m$ at $\mathbf{x}^{(i)}$ and $\nabla \Delta H(i, \mathbf{F}(\mathbf{X}))$ the m partial derivatives of the hypervolume contribution. Then*

$$\nabla \mathcal{H}_{\mathbf{F}}(i, \mathbf{X}) = \nabla \Delta H(i, \mathbf{F}(\mathbf{X})) \cdot \mathbf{f}'(\mathbf{x}^{(i)}), \ i = 1, \dots, \mu. \qquad (17)$$

Proof. This follows by rewriting Equation 16.   □

According to this new interpretation of Theorem 3, it can be said that *the i-th subvector of $\nabla \mathbf{F}(\mathbf{X})$ is the gradient of the hypervolume contributions at the i-th subvector of* $\mathbf{X}$ for all other values in $\mathbf{X}$ being constant.

Remark 5 (Visualization of 2-D and 3-D gradient). The fact that the components of the gradient are related to the gradients of the hypervolume contributions can be used for a graphical representation of the gradient of $\mathcal{H}$ at a μm-vector. For each subvector (that is for each point in the Pareto front approximation) the gradient vector is drawn as an arrow starting in that point. Normalization by dividing by the length of the subgradient, that is $||\nabla \Delta \mathcal{H}(i, \mathbf{Y})||$, makes the visualization more readable. Examples follows.

Example 4. The visualization in Figure 6 is based on the data of Example 3 and subvectors that contribute only zero gradient components are omitted.

Example 5. In Figure 7 a visualization for $m = 2$ and $\mu = 5$ is depicted. See Figure 8 for an example with $m = 3$ and with 100 points distributed randomly on the positive part of a sphere with radius 10 and a reference point of 0. Here normalization is used to make the picture more transparent.

Remark 6 (Implementation of 3-D Gradient). To implement the 3-D example in Figure 8 a fast computation of the 3-D Gradient field computation has been implemented in C++. It is based on the algorithm of Fonseca and Emmerich [27] that computes all contributions to the hypervolume indicator within a single sweep and with a time complexity in $\mathcal{O}(\mu \log \mu)$. This algorithm can be easily modified to compute the visible facets of the volumes that are dominated by precisely one single subvector and therewith the components of $\nabla \mathcal{H}$ at some μm vector within a *single* sweep. The details of this implementation are omitted in this paper, but the code is made available under `http://natcomp.liacs.nl`.

6.1   Optimality Conditions

From the theoretical observations in Theorem 6 necessary conditions for optimality of μd vectors w.r.t. the hypervolume indicator can be stated in a concise way.

Let us restrict our attention first to differentiable μd vectors $\mathbf{X}$ with all subvectors of $\mathbf{F}(\mathbf{X})$ being non-dominated and in the interior of $[\mathbf{r}, \infty)$. These μd vectors will be termed *proper μd-vectors*. Note, that for proper μd vectors all partial derivatives of $\mathbf{H}(\mathbf{F}(\mathbf{X}))$ are non-zero.

As $\mathbf{H}_F$ is differentiable in $\mathbf{X}$ the following optimality condition holds:

Theorem 7. *A necessary condition for $\mathcal{H}_F$ being optimal is that*

$$\nabla \Delta H(i, \mathbf{F}(\mathbf{X})) \cdot \mathbf{f}'(\mathbf{x}^{(i)}) = 0 \tag{18}$$

for all $i = 1, \ldots, \mu$, or in different notation

$$\sum_{k=1}^{m} \frac{\partial \Delta \mathcal{H}(i, \mathbf{F}(\mathbf{X}))}{\partial y_k^{(i)}} \cdot \frac{\partial f_k(\mathbf{x}^{(i)})}{\partial x_j^{(i)}} = 0 \tag{19}$$

for all $i = 1, \ldots, \mu; j = 1, \ldots, d$.

Proof. This is the usual condition for stationarity of points and decomposition of the gradient described in Theorem 6.   □

Loosely speaking, Theorem 7 means that by finding solutions for which all hypervolume contribution gradients turn zero, candidates for optimal approximation sets can be obtained. This observation yields μd equations to be satisfied for μd variables to be determined.

We note that this condition holds also for non-proper μd vectors, although we can already a-priori conclude that optima of non-proper μd-vectors are of minor interest. If our aim is to approximate an non-degenerate Pareto front, that is a $(m - 1)$-dimensional manifold, every one of the μ points in the approximation set should contribute.

A close look at Equation 18 shows that there can be two reasons that a proper μd-vector satisfies the equation for a particular index i:

1. Some components of the Jacobian matrix are zero.
2. The partial derivatives of the contribution (which remain constant for a fixed value of d) are canceled out by the components of the column vectors of the Jacobian matrix.

In the unconstrained bi-objective case the Fritz John necessary conditions for a differentiable point $\mathbf{x}$ (with respect to f_1 and f_2) to belong to the efficient set read:

$$\exists \lambda_1, \lambda_2 \geq 0 : \lambda_1 \neq \lambda_2 \text{ and } \lambda_1 \nabla f_1(\mathbf{x}) + \lambda_2 \nabla f_2(\mathbf{x}) = 0 \tag{20}$$

This means either at least one of the gradient vectors is zero, or the gradient vectors point in the opposite direction.

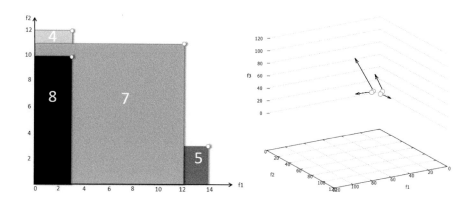

Fig. 6. Gradient computation for a 3-D gradient $\nabla\mathcal{H}(\mathbf{Y})$ for $\mathbf{Y} = ((3, 10, 8)\circ(12, 11, 7)\circ (14, 4, 5) \circ (6, 12, 4))^\top$ and reference point $\mathbf{r} = (0, 0, 0)^\top$

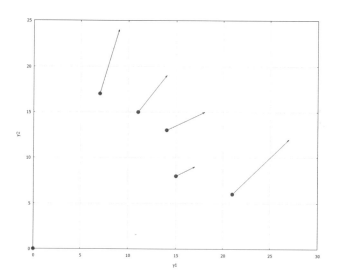

Fig. 7. Gradient computation for a 2-D gradient $\nabla\mathcal{H}(\mathbf{Y})$ for $\mathbf{Y} = ((7, 17) \circ (11, 15) \circ (14, 13) \circ (15, 8) \circ (21, 6))^\top$ and reference point $\mathbf{r} = (0, 0, 0)^\top$

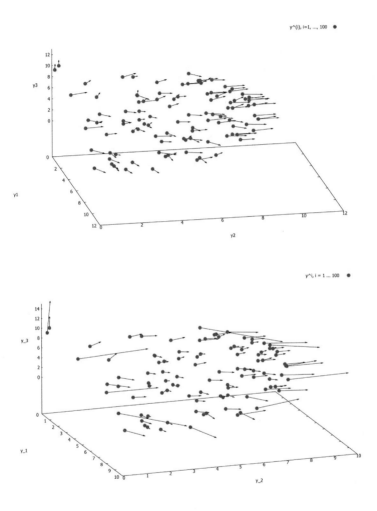

Fig. 8. Normalized gradient of $\mathcal{H}$ at **Y** (upper picture) and non-normalized gradient at **Y**, for **Y** given as a random set of objective vectors distributed randomly on the positive section of a sphere with radius 10

We can combine this with the previous result and obtain:

Corollary 1. *A necessary condition for a proper μd-vector consisting of efficient subvectors with respect to the bi-objective optimization problem to represent a locally optimal approximation set of the hypervolume indicator is given by:*

$$\frac{\frac{\partial \mathcal{H}}{\partial y_1^{(i)}}(\mathbf{F}(\mathbf{X}))}{\frac{\partial \mathcal{H}}{\partial y_2^{(i)}}(\mathbf{F}(\mathbf{X}))} = \frac{||\nabla f_2(\mathbf{x}^{(i)})||}{||\nabla f_1(\mathbf{x}^{(i)})||} \tag{21}$$

for all $i = 1, \ldots, \mu$.

In other words, the differential change in the i-th subvector of $\mathbf{X}$ in the decision space causes a growth of the hypervolume contribution of the i-th subvector of $\mathbf{F}(X)$ in the y_1 direction, that is compensated by a decrease of the hypervolume contribution of that subvector in direction y_2.

It is expected that a more general analysis of the findings presented in this section will reveal refined optimality conditions and a better understanding of the properties of (locally) optimal approximation sets. It may also shed new light on the yet unanswered question of how points in bounded size sets that maximize the hypervolume indicator distribute on a given Pareto front (see also [28]).

7   Conclusions and Outlook

This chapter refined the definition of the hypervolume indicator gradient field for the higher dimensional case. The size of the faces of the boundary of the dominated subspace are the gradient components at a set of objective vectors in the decision space. Partial derivatives of the gradient can be readily computed by using algorithms for computing the incremental hypervolume contributions. This yields algorithms with asymptotically optimal computational time complexity $\Theta(\mu d + \mu \log \mu)$ for computing the gradient at an approximation set from the Jacobian matrices of $\mathbf{f}$ at the points, and the values of the objective vectors in the bi- and tri-objective case. In the four objective case the time complexity can be guaranteed to be in $\mathcal{O}(\mu d + \mu^2)$. Further progress in incremental update schemes for the hypervolume indicator will also yield sharper bounds for gradient computations. Finally by deriving tight computation schemes, structural properties of the hypervolume indicator gradient field were revealed that entail a set of μd simple equations to be satisfied for an proper approximation set to be optimal. The analysis of these conditions may shed a light on the fundamental laws that govern the distribution of points in hypervolume indicator optimal approximations sets to the Pareto front (see also [28]). Moreover, the formulation of stopping criteria that guarantee local optimality for hypervolume-indicator based Pareto optimization is now in reach.

References

1. Emmerich, M.T.M., Deutz, A.H., Beume, N.: Gradient-Based/Evolutionary Relay Hybrid for Computing Pareto Front Approximations Maximizing the S-Metric. In: Bartz-Beielstein, T., Blesa Aguilera, M.J., Blum, C., Naujoks, B., Roli, A., Rudolph, G., Sampels, M. (eds.) HCI/ICCV 2007. LNCS, vol. 4771, pp. 140–156. Springer, Heidelberg (2007)

2. Fliege, J., Svaiter, B.F.: Steepest Descent Methods for Multicriteria Optimization. Mathematical Methods of Operations Research 51(3), 479–494 (2000)

3. Brown, M., Smith, R.E.: Effective Use of Directional Information in Multi-objective Evolutionary Computation. In: Cantú-Paz, E., et al. (eds.) GECCO 2003. LNCS, vol. 2723, pp. 778–789. Springer, Heidelberg (2003)

4. Bosman, P.A., de Jong, E.D.: Exploiting Gradient Information in Numerical Multi-Objective Evolutionary Optimization. In: Beyer, H.G., et al. (eds.) GECCO 2005, vol. 1, pp. 755–762. ACM Press, New York (2005)

5. Lara, A., Schütze, O., Coello, C.A.C.: On Gradient-Based Local Search to Hybridize Multi-objective Evolutionary Algorithms. In: Tantar, E., Tantar, A.-A., Bouvry, P., Del Moral, P., Legrand, P., Coello Coello, C.A., Schütze, O. (eds.) EVOLVE- A bridge between Probability, Set Oriented Numerics and Evolutionary Computation. SCI, vol. 447, pp. 303–330. Springer, Heidelberg (2013)

6. Timmel, G.: Ein stochastisches Suchverfahren zur Bestimmung der Optimalen Kompromißlösungen bei statistischen polykriteriellen Optimierungsaufgaben. Journal TH Ilmenau 6, 139–148 (1980)

7. Schäffler, S., Schultz, R., Wienzierl, K.: Stochastic Method for the Solution of Unconstrained Vector Optimization Problems. Journal of Optimization Theory and Applications 114(1), 209–222 (2002)

8. Shukla, P.K., Deb, K., Tiwari, S.: Comparing Classical Generating Methods with an Evolutionary Multi-objective Optimization Method. In: Coello Coello, C.A., Hernández Aguirre, A., Zitzler, E. (eds.) EMO 2005. LNCS, vol. 3410, pp. 311–325. Springer, Heidelberg (2005)

9. Hillermeier, C.: Generalized Homotopy Approach to Multiobjective Optimization. Journal of Optimization Theory and Applications 110(3), 557–583 (2001)

10. Schütze, O., Dell'Aere, A., Dellnitz, M.: Continuation Methods for the Numerical Treatment of Multi-Objective Optimization Problems. In: Branke, J., Deb, K., Miettinen, K., Steuer, R. (eds.) Practical Approaches to Multi-Objective Optimization. Dagstuhl Seminar Proceedings, vol. 04461. IBFI, Schloss Dagstuhl, Germany (2005)

11. Schütze, O., Lara, A., Coello Coello, C.A.: The Directed Search Method for Unconstrained Multi-Objective Optimization Problems. In: Proceedings of the EVOLVE–A Bridge Between Probability, Set Oriented Numerics, and Evolutionary Computation (2011)

12. Ehrgott, M.: Multicriteria Optimization. Springer (2005)

13. Zitzler, E., Thiele, L.: Multiobjective Optimization Using Evolutionary Algorithms—A Comparative Case Study. In: Eiben, A.E., Bäck, T., Schoenauer, M., Schwefel, H.-P. (eds.) PPSN 1998. LNCS, vol. 1498, pp. 292–301. Springer, Heidelberg (1998)

14. Zitzler, E., Thiele, L., Laumanns, M., Fonseca, C.M., da Fonseca, V.G.: Performance Assessment of Multiobjective Optimizers: an Analysis and Review. IEEE Trans. Evolutionary Computation 7(2), 117–132 (2003)

15. Auger, A., Bader, J., Brockhoff, D., Zitzler, E.: Hypervolume-based Multiobjective Optimization: Theoretical Foundations and Practical Implications. Theor. Comput. Sci. 425, 75–103 (2012)
16. Beume, N.: Hypervolume-Based Metaheuristics for Multiobjective Optimization. PhD Thesis. Eldorado (2011)
17. Custódio, A.L., Emmerich, M., Madeira, J.F.A.: Recent Developments in Derivative-free Multiobjective Optimization. In: Topping, B. (ed.) Computational Technology Reviews, vol. 5, pp. 1–30. Saxe-Coburg Publications (2012)
18. Bringmann, K.: Bringing Order to Special Cases of Klee's Measure Problem. CoRR abs/1301.7154 (2013)
19. Beume, N., Fonseca, C.M., López-Ibáñez, M., Paquete, L., Vahrenhold, J.: On the Complexity of Computing the Hypervolume Indicator. IEEE Trans. Evolutionary Computation 13(5), 1075–1082 (2009)
20. Yıldız, H., Suri, S.: On Klee's Measure Problem for Grounded Boxes. In: Dey, T.K., Whitesides, S. (eds.) Symposium on Computational Geometry, pp. 111–120. ACM (2012)
21. Fonseca, C.M., Guerreiro, A.P., López-Ibáñez, M., Paquete, L.: On the Computation of the Empirical Attainment Function. In: [29], pp. 106–120
22. Zitzler, E., Thiele, L., Laumanns, M., Fonseca, C.M., Grunert da Fonseca, V.: Performance Assessment of Multiobjective Optimizers: An Analysis and Review. IEEE TEC 7(2), 117–132 (2003)
23. Guerreiro, A.P., Fonseca, C.M., Emmerich, M.T.M.: A Fast Dimension-Sweep Algorithm for the Hypervolume Indicator in Four Dimensions. In: CCCG, pp. 77–82 (2012)
24. Kung, H.T., Luccio, F., Preparata, F.P.: On Finding the Maxima of a Set of Vectors. Journal of the ACM 22(4), 469–476 (1975)
25. Baeza-Yates, R.: A Fast Set Intersection Algorithm for Sorted Sequences. In: Sahinalp, S.C., Muthukrishnan, S.M., Dogrusoz, U. (eds.) CPM 2004. LNCS, vol. 3109, pp. 400–408. Springer, Heidelberg (2004)
26. Hupkens, I., Emmerich, M.: Logarithmic-time Updates in SMS-EMOA and Hypervolume-based Archiving. In: Emmerich, M., et al. (eds.) EVOLVE - A Bridge between Probability, Set Oriented Numerics,and Evolutionary Computation IV. AISC, vol. 227, pp. 155–169. Springer, Heidelberg (2013)
27. Emmerich, M.T.M., Fonseca, C.M.: Computing Hypervolume Contributions in Low Dimensions: Asymptotically Optimal Algorithm and Complexity Results. In: [27], pp. 121–135
28. Auger, A., Bader, J., Brockhoff, D., Zitzler, E.: Theory of the Hypervolume Indicator: Optimal μ-Distributions and the Choice of the Reference Point. In: Foundations of Genetic Algorithms (FOGA 2009), pp. 87–102. ACM, New York (2009)
29. Takahashi, R.H.C., Deb, K., Wanner, E.F., Greco, S. (eds.): EMO 2011. LNCS, vol. 6576. Springer, Heidelberg (2011)

A Multi-Directional Modified Physarum Algorithm for Optimal Multi-Objective Discrete Decision Making

L. Masi and M. Vasile

Department of Mechanical & Aerospace Engineering, University of Strathclyde, 75 Montrose Street, G1 1XJ, Glasgow, UK
{luca.masi,massimiliano.vasile}@strath.ac.uk

Abstract. This paper will address an innovative bio-inspired algorithm able to incrementally grow decision graphs in multiple directions for discrete multi-objective optimisation. The algorithm takes inspiration from the slime mould *Physarum Polycephalum*, an amoeboid organism that in its plasmodium state extends and optimizes a net of veins looking for food. The algorithm is here used to solve multi-objective Traveling Salesman and Vehicle Routing Problems selected as representative examples of multi-objective discrete decision making problems. Simulations on selected test cases showed that building decision sequences in two directions and adding a matching ability (multi-directional approach) is an advantageous choice if compared with the choice of building decision sequences in only one direction (unidirectional approach). The ability to evaluate decisions from multiple directions enhances the performance of the solver in the construction and selection of optimal decision sequences.

Keywords: Physarum, Multi-Objective, Multi-Directional.

1 Introduction

The idea that nature can inspire humans to solve complex decision making problems was widely used over the past two decades. A number of bio-inspired algorithms designed to solve decision making problems was and still are studied and developed. An example is the ACO (Ant Colony Optimisation) alghorithm [1] that takes inspiration from the social behaviour of ants looking for food. Following this concept, the behaviour of other social animals was successfully used: examples are bee colonies [2], fireflies [3], birds [4].

The method proposed in this paper takes inspiration from *Physarum Polycephalum*, see Fig. 1, known as many-headed slime mould, a simple organism inahabiting moist areas that was endowed by nature with heuristics that can be used to solve single-objective and multi-objective discrete decision making problems. In [5] it has been shown that *Physarum Polycephalum* is able to solve a maze finding the shortest path that connects the maze's entrance and exit by changing its shape. It has been shown also that a living *Physarum* is able to recreate the Japan rail network [6] and the Mexican highway network [7], both

O. Schütze et al. (eds.), *EVOLVE - A Bridge between Probability, Set Oriented Numerics, and Evolutionary Computation III*, Studies in Computational Intelligence 500,
DOI: 10.1007/978-3-319-01460-9_9, © Springer International Publishing Switzerland 2014

Fig. 1. Physarum Polycephalum. Image courtesy of *Howard County Bird Club* at www.howardbirds.org.

using an experimental arena with food sources at each of the major cities in the regions. *Physarum* based algorithms have been developed recently to solve multi-source problems with a simple geometry [8,9], mazes [10] and transport network problems [6,10].

In this paper a multi-directional modified *Physarum Polycephalum* algorithm able to solve NP-hard multi-objective classical problems in operations research is proposed. In [11,12,13] multi-objective bio-inspired algorithms, i.e. ant colony algorithms, have been proposed and studied. The algorithm presented in this work is a multi-objective generalization of the single-objective multi-directional modified *Physarum* solver previously presented in [14] for discrete decision making.

In Sect. 2 the physiology of *Physarum* is introduced: discrete decision making problems are modeled with decision graphs where nodes represent the possible decisions while arcs represent the cost vector associated with decisions. Each arc has a scalar dominance index associated which is calculated comparing all the arcs leaving a node, as explained in Sect. 2. Decision graphs are incrementally grown and explored in multiple directions using the *Physarum*-based heuristic. This paper aims at proving that a multi-directional incremental *Physarum* solver is more efficient, in terms of success indexes (see Sect. 3.1), than a unidirectional incremental *Physarum* solver when applied to the solution of multi-objective decision problems that can be represented with directed symmetric decision graphs, i.e. graphs where the contribution of an arc to a complete path can be evaluated moving forward or backward along the graph. In [14] it has been already shown that the single-objective multi-directional modified *Physarum* algorithm is more efficient than the unidirectional algorithm when applied to small scale single-objective discrete decision making problems. This thesis will be demonstrated for the multi-objective algorithm in Sect. 4 solving some test cases.

Bi-objective Traveling Salesman and Vehicle Routing Problems (TSP and VRP), introduced in Sect. 3, with a number of nodes between 10 and 100, were chosen as representative examples of the above type of decision making problems, here called reversible decision-making problems, i.e. problems in which a decision can be taken either moving forward or backward along the graph, as explained in Sect. 2.

2   Biology and Mathematical Modeling

Physarum Polycephalum is a large, single-celled amoeboid organism that exhibits intelligent plant-like and animal-like characteristics. Its main vegetative state, the *plasmodium*, is formed of a network of veins (*pseudopodia*). The stream in these tubes is both a carrier of chemical and physical signals, and a supply network of nutrients throughout the organism [9]. *Physarum* searches for food by extending this net of veins, whose flux is incremented or decremented depending on the food position with reference to its centre. The longest is the path connecting the centre with the source of food, the smallest is the flux and viceversa: best veins in terms of length that connect its centre with the food tend to increase their radius and the flux of nutrients inside, while longer veins tend to decrement the flux and close with time. This behaviour can be interpreted as a natural attitude in optimising the energy required to feed the organism by shape variation.

2.1   Problem Formulation: Multi-Objective Discrete Decision Making

Given a solution j to a discrete multi-objective decision making problem P, with cost vector $\mathbf{s}^j = [s_1^j, s_2^j, ..., s_n^j]$ and a solution i with cost vector $\mathbf{s}^i = [s_1^i, s_2^i, ..., s_n^i]$, the solution j dominates i if $s_k^j \leq s_k^i$ for all the $k = 1, 2, ..., n$ and $s_k^j < s_k^i$ for at least one k. The relation $\mathbf{s}^j \prec \mathbf{s}^i$ states that $\mathbf{s}^j$ dominates $\mathbf{s}^i$. The dimension of the vector $\mathbf{s}^j$ expresses the number of evaluating criteria for a solution j. The cost vector represents the cost associated with a decision. A general problem in discrete multi-objective optimisation is to find the feasible non dominated solutions to the given discrete multi-objective decision making problem P. Following the theory developed in [18], it is possible to associate a scalar dominance index $I(\mathbf{s})$ to each solution. The lower is the index, the better is the solution: if one considers the set of solutions $S = \{\mathbf{s}^j, \mathbf{s}^i, \mathbf{s}^k\}$ where $\mathbf{s}^j \prec \mathbf{s}^i \prec \mathbf{s}^k$, the set of associated scalar indexes will be $I = \{I(\mathbf{s}^j) = 0, I(\mathbf{s}^i) = 1, I(\mathbf{s}^k) = 2\}$. All the non-dominated solutions in a general set S form the set:

$$PF = \{\mathbf{s}|I(\mathbf{s}) = 0\} \tag{1}$$

which is called Pareto front. Therefore, the solution of the problem P translates into finding the elements of PF.

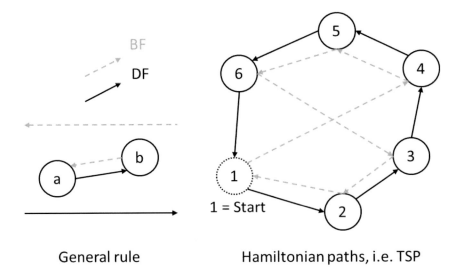

Fig. 2. Generally a *Physarum* working in direct flow (*DF*) would build the decision that brings from a to b, while the *Physarum* working in back flow (*BF*) would build the symmetric decision bringing from b to a (left). In a traveling salesman problem (TSP, right) nodes are fixed while arcs are built with time and a decision that brings from a to b can be built from both DF and BF *Physarum*, as for the arc that connect 4 to 5.

2.2   Multi-Objective Multi-Directional *Physarum* Algorithmic

A reversible discrete decision problem can be modeled using a symmetric directed graph. The reversibility of a decision that induces a change from a state a to state b indicates here that the decision that brings back from b to a exists and can be evaluated. Not necessarily these two decisions have the same cost. The symmetric directed graph can be seen as the superposition of two directed graphs (direct-flow, *DF*, and back-flow, *BF*, graphs) whose nodes are coincident and edges have opposite orientation. In so doing, the decision between state a and b has a forward link a to b and a superposed backward link b to a. It is assumed that the first decision node is the heart of a growing *Physarum* in *DF*, and the end decision node the heart of a growing *Physarum* in *BF*. The two *Physarum* are supposed able to incrementally grow the decision graph in the two directions by extending their net of veins. A multiple direction growing decision *Physarum* graph is obtained. In the example mentioned before, the *Physarum* working in DF would build its graph by creating arcs that move from a to b. If the graph was traversed by a virtual agent, the agent would walk along an arc from a to b. The other *Physarum* would build its graph in the opposite direction then walking along each arc from b to a. The result is a graph where both nodes and links are incrementally built by two expanding *Physarum*.

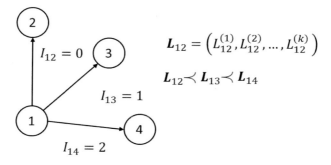

Fig. 3. Example of acrs' values assignment. $\boldsymbol{L}_{ij}$ and I_{ij} are respectively the cost vector and dominance index associated with the general decision from node i to node j.

In this work the decisional problems analyzed are the TSP and the VRP, see Sect. 3; being Hamiltonian paths, all the nodes are built at the beginning, and only the arcs are built with time. This means that both the *Physarum* working in DF and the one working in BF can build arcs connecting two nodes in both directions. If the graphs built by the two *Physarum* are fully connected after a transient of growth, which is the case in this paper, two superposed symmetric directed graph are obtained. Fig. 2 shows a simple example for a TSP problem.

Exploration Using the Hagen-Poiseuille law, the flux through the net of *Physarum* veins is [6,8,9,10]:

$$Q_{ij} = \frac{\pi r_{ij}^4}{8\mu} \frac{\Delta p_{ij}}{L_{ij}} \qquad (2)$$

where Q_{ij} is the flux between i and j, μ is the dynamic viscosity, r_{ij} the radius, L_{ij} the length and Δp_{ij} the pressure gradient.

In a multi-objective algorithm the length L_{ij}, representing the cost of the decision that brings from i to j, is a vector $\boldsymbol{L}_{ij}$. Its value can be substituted with the scalar dominance index $I_{ij} \in \mathbb{N}^0$, see Fig. 3: the cost vectors associated with each veins that connect a node N_i with other nodes N_j^k can be compared and the dominance indexes can be evaluated. Eq. (2) becomes:

$$Q_{ij} = \frac{\pi r_{ij}^4}{8\mu} \frac{\Delta p_{ij}}{(I_{ij} + 1)} \qquad (3)$$

where plus 1 was added to avoid a singularity for $I_{ij} = 0$.

The strategy of using a single structure (in this case the flux) has been previously examined in [11] for Ant Colony Optimisation applied to multi-objective problems (where a weighted sum of all the objectives is used instead of the index I_{ij}). Another strategy, as reported in [11], could be the use of several fluxes structures ([11] refers to pheromones), one for each objective. The first strategy was chosen because it has the advantage of being easy to implement and usable when the number of objectives is high, as in many real-world problems. This considerations will be further discussed in Sect. 4.1.

Table 1. Input parameters for the modified *Physarum* solver

m	Linear dilation coefficient, see Eq. (9).
ρ	Evaporation coefficient, see Eq. (5).
GF_{ini}	Initial growth factor, see Eq. (7)
N_{agents}	Number of virtual agents.
p_{ram}	Probability of ramification, see Sect. 2.2.
α	Weights on ramification, see Eq. (8).
$k_{explosion}$	radii upper limit, see Eq. (10)

Diameter variations then cause a change in the flux. Veins' dilation due to an increasing number of nutrients flowing can be modeled using a monotonic function of the flux:

$$\frac{d}{dt}r_{ij}\bigg|_{dilation} = f(Q_{ij}) \tag{4}$$

where $f(0) = 0$, i.e. linear, sigmoidal, etc. Veins' contraction, similarly to the evaporative effect in ACO [1], can be assumed to be linear with radius:

$$\frac{d}{dt}r_{ij}\bigg|_{contraction} = -\rho r_{ij} \tag{5}$$

where $\rho \in [0,1]$ is defined evaporation coefficient. The probability associated with each vein connecting i and j is then computed using a simple adjacency probability matrix based on fluxes:

$$P_{ij} = \begin{cases} \frac{Q_{ij}}{\sum_{j \in N_i} Q_{ij}} & if\ j \in N_i \\ 0 & if\ j \notin N_i \end{cases} \tag{6}$$

where N_i is the set of neighbour for i.

An additive term in the veins' dilation process, whose first main term is expressed in Eq. (4) was added in the algorithm and takes inspiration from the behaviour of the amoeba *Dictyostelium discoideum* [16]. This dilation is:

$$\frac{d}{dt}r_{ij_{best}}\bigg|_{elasticity} = GF r_{ij_{best}} \tag{7}$$

where GF is the growth factor and $r_{ij_{best}}$ the veins' radius of the best chains of veins, i.e. the veins that form the paths in the decision graph that are in the current calculated Pareto front. This dilation, as explained in [14], simulates the tendency of best veins to further increase their radius for the effect of the flux.

Growth in Multiple Directions and Matching. The incremental growth of decision network in multiple directions is then based on a weighted roulette. Nutrients inside veins are interpreted as virtual agents that move in accord with adjacency probability matrix in Eq.(6). Once a node is selected, there is a probability p_{ram} of ramification towards new nodes that are not yet connected with the actual node. The value of p_{ram} can be chosen *a priori* or a law can be defined, i.e. $p_{ram}^c = p_{ram}^c(A_c)$, where c is the current ramifying node and A_c the number of arcs that leave node c. In this work *a priori* values were chosen before the simulations.

If ramification is the choice, a weighted roulette, based on objective function evaluations, helps the *Physarum* with the selection and construction of a new link. The probability of a new link construction from the current node c to a new possible node $n_i \in N$, where N is the set of new possible decisions, is here assumed to be inversely proportional to the cost I_{cn_i} of the decision between c and n_i, i.e. the dominance scalar index associated with the decision:

$$p_{cn_i} \propto \frac{1}{(I_{cn_i} + 1)^\alpha} \tag{8}$$

where α is a weight. Once a new link is built, a complete decision path is constructed (creating other links if necessary).

Assuming then the presence of two counter expanding *Physarum*, one in direct-flow *DF* and one superposed in back-flow *BF*, as explained in the previous paragraph, a matching condition can be then defined. If an arc connecting two nodes that belongs to *DF* and *BF Physarum* respectively, exists or can be created, it is traversed by the agent and becomes part of both the *DF* and the *BF*. Some matching strategies were compared in [14] for single-objective problems. In this paper, two matching strategies were implemented in the multi-objective modified *Physarum* solver.

The first one, called *selective-matching*, follows an elitist criterion where at each generation a joint path is selected if and only if its total cost vector is not dominated by the previous joint paths selected during the same generation, as in [14]. It could be noted that if a high number of exploring agents is chosen, a high number of paths are matched. This could lead to a slowdown of the code speed, especially if complete decision sequences are long, as in more complex multi-objective VRPs and TSPs (more than 20 cities). For this reason, an other strategy, called *mix-matching*, was designed for larger scale problems. Selective matching is done only considering the best n solutions in DF and BF during a generation, so that worst routes are excluded *a priori*. A value $n = \frac{dim}{5}$, where *dim* is the problem dimension, i.e. the number of cities for TSP and VRP, was used in the simulations presented in this paper. Furthermore the n best decisions in DF and BF are matched with the Pareto front found by the algorithm at the time of matching.

These matching conditions can be interpreted as a communication ability between the two *Physarum*: they move according to their nature and the knowledge acquired exploring the decision space, which contains both personal experience and shared information.

Restart Procedure. Simulations on selected test cases (see Sect. 4) were carried out adding in the *Physarum* algorithm two restart procedures to avoid stagnation on local minima. The first restart procedure, called *restart1*, is a routine for the adaptive control of the growth factor GF. This control was introduced in order

---

Algorithm 1. Multi-directional incremental modified Physarum solver

---

initialize m, ρ, GF_{ini}, N_{agents}, p_{ram}, α
generate a random route from start to destination both in DF and BF
for each generation **do**
   for each virtual agent in all directions (DF and BF) **do**
      if current node $\neq$ end node **then**
         if $rand \leq p_{ram}$ **then**
            using Eq. (8) create a new link to a node not yet connected
            update scalar dominance indexes for current node, see Sect. 2.1
         else
            move on existing graph using Eq. (6)
         end if
      end if
   end for
   look for possible matchings among decision sequences in DF and BF
   update Pareto front
   contract and dilate veins using Eqs. (4), (5), (7)
   if r_{ij} exceeds upper radius limit, see Eq. (10) **then**
      block radius increment
   end if
   update fluxes and probabilities using Eqs. (3), (6)
end for

---

to incrementally boost the effect of GF during a simulation, driving exploring agents towards best veins. Simulations showed that the adaptive control of GF helps the convergence of the algorithm towards optimal solution when used on small scale problems (number of cities less than 16 in this paper). Given an initial value for the growth factor GF_{ini}, GF is incremented by a fixed percentage σ after every generation. If the highest probability p_{best}^{PF} associated with the paths in the calculated Pareto front so far is higher than a fixed value p_{lim}^{low}, the increment is set to zero. Then, if p_{best}^{PF} exceeds a value p_{lim}^{high}, GF is set equal to GF_{ini} and veins are dilated and contracted to their initial value. In the present paper is assumed $\sigma = 0.01$, $p_{lim}^{low} = 10^{-4}$, $p_{lim}^{high} = 0.85$ for the bi-objective Vehicle Routing Problem test case *Tuscany10* and $p_{lim}^{high} = 0.95$ for the bi-objective Traveling Salesman Problem test case *Ulysses16*. A second restart procedure, called *restart2*, was designed for larger scale problem, i.e. the bi-objective traveling salesman instance *KroAB100*. It is based on minimum nodes in common among decision sequences in a generation and among decision sequences and Pareto front; the algorithm is restarted if one of the following conditions is achieved:

I) the minimum number of nodes in common n_{min}^{com}, obtained comparing all decision sequences among each other in a generation, exceeds a threshold n_{com}.

Table 2. Values used as input parameters - TSP test case (first row) and VRP test case (second row)

Instance	m	ρ	GF_{ini}	N_{agents}	p_{ram}	α	$k_{explosion}$
Ulysses16	5×10^{-5}	$\frac{1 \cdot 10^{-5}}{N_{agents}}$	$5 \cdot 10^{-3}$	100	0.8	0	10^8
Tuscany10	5×10^{-5}	$\frac{1 \cdot 10^{-5}}{N_{agents}}$	$1 \cdot 10^{-3}$	150	0.8	0	10^8
KroAB100	5×10^{-5}	$\frac{5 \cdot 10^{-6}}{N_{agents}}$	$5 \cdot 10^{-3}$	50	1	0	5

II) a fraction β of the decision sequences built during a generation belong to the calculated Pareto front at same generation.

In this paper a value $n_{com} = \frac{dim}{2}$, where dim is dimension of the problem, i.e. the number of cities, and a value $\beta = \frac{2}{3}$ were used. The goal of this restart procedure is to avoid both a stagnation to local single minima and to the calculated Pareto front itself.

Considerations on the Algorithm. The set of Eqs. (2)-(6) can be implemented as in the following. In accordance to Eq. (2), flux in each vein is proportional to the radius and inversely proportional to the length (the scalar dominance index in a multi-objective problem, Eq. (3)). These two main parameters are taken into account in the algorithm. Once a vein is selected by a virtual agent in a generation, its radius is incremented using Eq. (4). In the present work, a function linear with respect to the product between the radius $r_{ij}^{(n)}$ of the veins traversed by agent n, and the inverse of the sum of dominance indexes ($I_{tot}^{(n)}$, see Sect. 2.1), associated with each arc of the decision taken by agent n, will be used for the veins' dilation:

$$\left. \frac{d}{dt} r_{ij}^{(n)} \right|_{dilation} = m \frac{r_{ij}^{(n)}}{I_{tot}^{(n)}} \qquad (9)$$

where the coefficient m is the linear dilation coefficient. Evaporation is taken into account using Eq. (5) for each agent. Fluxes are then calculated using Eq. (3) and probabilities are updated in accordance with Eq. (6). Due to the mathematical nature of the algorithm (the flux is related to the fourth power of the radius), an upper limit on the maximum vein radius was introduced in order to avoid veins' flux explosion. If the radius r_{ij} exceeds a maximum value r_{max}, the vein dilation is blocked up until the radius is again below r_{max} for the effect of evaporation. This upper limit, called $k_{explosion}$, is given as ratio between r_{ij} and r_{ini}:

$$k_{explosion} = \frac{r_{ij}}{r_{ini}} \qquad (10)$$

where r_{ini} is the initial radius of the veins. The main parameters of the modified *Physarum* solver are listed in Table 1. The initial radius of the veins r_{ini} is always set equal to 1 in the simulations presented in this paper. The pseudocode of the multi-directional incremental modified *Physarum* solver is provided in Algorithm 1.

3 Application to Multi-Objective Traveling Salesman and Vehicle Routing Problems and Benchmark

In single-objective optimisation the Traveling Salesman problem, TSP, is the problem of finding the shortest tour that visit each city of a given set S of n cities.

In the multi-objective optimisation case considered in this paper the cost function to be minimized is a vector of two values: the total length $L_{tot} = \sum_j L_j$ and the total road traffic $T_{tot} = \sum_j T_j$ of each tours, where $j = 1, ..., n$ is the index that identifies each part of the tour. The road traffic is here assumed to be inversely proportional to the length $T_j = 1/L_j$. The shorter is the tour, the higher is the probability that the tour is chosen by drivers, increasing the road traffic. The total road traffic $T_{tot} = \sum_j T_j$ will be called Road Traffic Index in the following. Although conflicting criteria, both length and road traffic in a tour have to be minimised.

TSPLIB [17] was used to benchmark the proposed *Physarum* algorithm, developed in Matlab® R2010b, on the TSP problem. In Sect. 4 are reported the results obtained by applying the multi-objective multi-directional *Physarum* solver to test case *Ulysses16* that was modified adding the road traffic to the cost function and to the test case *KroAB100*, obtained from the single-objective intances *KroA100* and *KroB100*. For the *KroAB100* the consideration above on the road traffic index does not apply: it is a bi-objective problem itself and two objectives to be minimised are included in the instance.

The multi-objective Vehicle Routing Problem, VRP, considered in this paper is a similar problem. Given a set of n cities with a demand k, whose reciprocal distance L_j and road traffic $T_j = 1/L_j$ are known, v vehicles of capacity c, d depots located in fixed cities, the VRP is the problem of delivering goods located in the depots using a defined amount of vehicles with finite capacity. The goal is to satisfy the demand of each city minimizing the cost functions, i.e. the distance and road traffic. VRP reduces to a TSP if there is only one vehicle with infinite capacity. When the modified *Physarum* algorithm is applied to VRP, a probability skew factor ψ is included in the algorithm. If an agent is not obliged to go to depot, the probability to reach the depot is lowered of a factor $(1 - \psi)$. Other probabilities are then risen of a same value in order to have the sum of probabilities equals to 1. The skew factor ψ is introduced in the model to avoid frequent returns to depot in the decision sequences and is here set equals to 0.5. The *Physarum* solver applied to VRP was tested on a map of 9 cities plus one depot. The map is built using 9 Italian cities (Firenze, Livorno, Montecatini, Pistoia, Prato, Montevarchi, Arezzo, Siena, San Gimignano), with a city considered the depot (Ponsacco). The Euclidean distance in kilometers was used. VRP parameters were set to $n = 9, k = cost = 1, v = 1, c = 4, d = 1$, i.e. one vehicle with capacity equals to 4, one depot and a constant demand equals to 1.

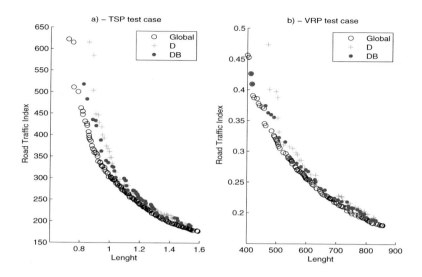

Fig. 4. Pareto front - TSP test case *Ulysses16* at $6.5 \cdot 10^5$ function evaluations, (a) - VRP test case *Tuscany10* at $3 \cdot 10^5$ function evaluations, (b). In the legend, *Global* indicates the global Pareto front obtained from all the runs of the D and D&B algorithms (1600 for the VRP test case and 2800 for the TSP test case), while *D* and *D&B* indicate an example of a Pareto front found during a run of the D and D&B algorithms respectively.

3.1 Testing Procedure

The testing procedure proposed in [18] was used in this paper. Two metrics are defined:

$$M_{spr} = \frac{1}{M_p} \sum_{i=1}^{Mp} \min_{j \in N_p} \| \frac{\mathbf{f}_j - \mathbf{g}_i}{\mathbf{g}_i} \| \tag{11}$$

$$M_{conv} = \frac{1}{N_p} \sum_{i=1}^{Np} \min_{j \in M_p} \| \frac{\mathbf{g}_j - \mathbf{f}_i}{\mathbf{g}_j} \| \tag{12}$$

where M_p is the number of elements, with objective cost function $\mathbf{g}$, in the true global Pareto front and N_p is the number of elements, with objective cost function $\mathbf{f}$, in the Pareto front that a given algorithm is producing. Although similar, the two metrics are measuring two different things: M_{spr} is the sum, over all the elements in the global Pareto front, of the minimum distance of all the elements in the Pareto front N_p from the ith element in the global Pareto front: this metric would be high if N_p was only a partial representation of the global Pareto front. M_{conv}, instead, is the sum, over all the elements in the Pareto front N_p, of the minimum distance of the elements in the global Pareto front from the ith element in the Pareto front N_p: this metric would give a low

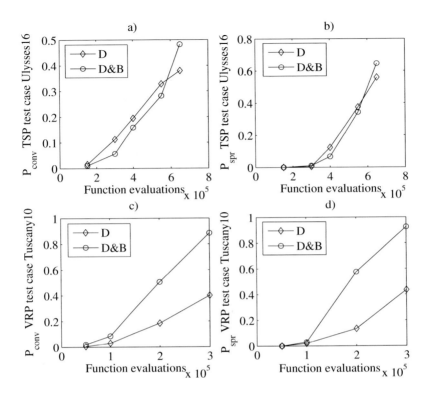

Fig. 5. Variation of the indexes of performance p_{conv} and p_{spr} with the number of function evaluations - TSP test case *Ulysses16*, (a) and (b) - VRP test case *Tuscany10*, (c) and (d)

value if N_p was an accurate, although partial, representation of the global Pareto front.

From the considerations above, both the metrics M_{spr} and M_{conv} should be low for a good estimate of the global calculated Pareto front. The indexes of performance $p_{conv} = P(M_{conv} < tol_{conv})$ and $p_{spr} = P(M_{spr} < tol_{spr})$ will be used to explore the efficiency of the algorithm and to compare the multi-directional and the unidirectional versions. Given n repeated runs, p_{conv} is the probability that M_{conv} achieves a value less than tol_{conv}, while p_{spr} is the probability that M_{spr} achieves a value less than tol_{spr}. 200 runs are sufficient in order to obtain an error $\leq 5\%$ with a 95% of confidence [18]. For the TSP test case *Ulysses16* the tolerances tol_{conv} and tol_{spr} are set equal to 0.0465 and 0.045 respectively, for the VRP test case *Tuscany10* to 0.030 and 0.035, and for the TSP test case *KroAB100* to 0.048 and 0.058.

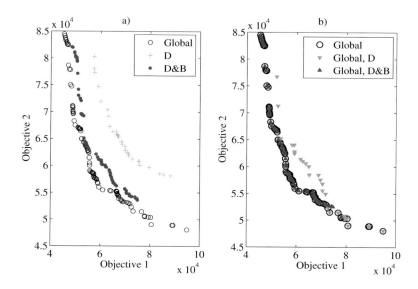

Fig. 6. Pareto fronts for TSP test case *KroAB100* at $4 \cdot 10^7$ function evaluations. In the legend of figure a), *Global* indicates the global Pareto front obtained from all the runs of the D and D&B algorithms (40), while *D* and *D&B* indicate an example of a Pareto front found during a run of the D and D&B algorithms respectively. In the legend of figure b), *Global* indicates the global Pareto front as in a), while *Global, D* and *Global, D&B* indicate the global Pareto fronts obtained from all the runs of the D and D&B algorithms respectively (20).

4 Results

The multi-objective multi-directional modified *Physarum* solver, named D&B in the following, was compared against a multi-objective unidirectional modified *Physarum* solver, named D. The D algorithm is obtained by freezing the backflow BF. The two algorithms were applied to the modified symmetric traveling salesman problem test case *Ulysses16*, to the symmetric traveling salesman problem test case *KroAB100* and to the vehicle routing problem test case *Tuscany10*, described in Sect. 3. The values used as input parameters in the simulations, chosen after a series of trials, are listed in Table 2. Selected ones showed best performance. The restart procedure *restart1* and the matching strategy *selective-matching* were used for *Ulysses16* and *Tuscany10*, while *restart2* and *mix-matching* were used for *KroAB100*. Simulations were carried out on a 64-bit OS Windows 7 Intel® Core™2 Duo CPU E8500 3.16GHz 3.17GHz.

Ulysses16 & Tuscany10. In both the test cases the true global Pareto front was unknown. In order to obtain a global Pareto front all the runs (1600 for the VRP test case *Tuscany10* and 2800 for the TSP test case *Ulysses16*) of the multi-directional and unidirectional algorithms were used: two global Pareto fronts for

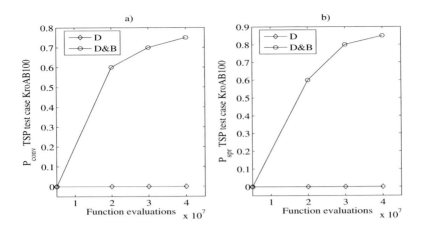

Fig. 7. Variation of the indexes of performance p_{conv} and p_{spr} with the number of function evaluations - TSP test case *KroAB100*

both the VRP and TSP test cases were built using all the solutions found by the algorithms. In Fig. 4 the global Pareto fronts are shown. The figure reports also an example of Pareto front found by unidirectional (D) and multi-directional (DB) algorithms, for both *Ulysses16* and *Tuscany10*. Fig. 5 shows the variation of the indexes of performance p_{conv} and p_{spr} with the number of function evaluations for the TSP test case ((a) and (b)) and for the VRP test case ((c) and (d)). A function evaluation is defined as the call to the objective function, i.e. each arc selected by the virtual exploring agents (see Sect. 2.2) is considered a function evaluation. Results for the VRP test case *Tuscany10*, as shown in Fig. 5 (c) and (d), demonstrate that the multi-objective multi-directional modified *Physarum* algorithm with matching ability (D&B) provides higher indexes of performance p_{conv} and p_{spr}, than the multi-objective unidirectional modified *Physarum* algorithm (D), at all the function evaluations limit. This gain is up to approximately 50% for the p_{spr} and p_{conv} at $3 \cdot 10^5$. The results for TSP test case *Ulysses16*, reported in Fig. 5 (a) and (b), show that the multi-directional algorithm provides better performance after $6 \cdot 10^5$ function evaluations and the gain is up to 10% for both p_{spr} and p_{conv} at $6.5 \cdot 10^5$. The behaviour of the indexes of performance for this multi-objective instance are similar to the behaviour of the index of performance in [14] for the same TSP test case with single-objective: the unidirectional algorithm tends to have a better performance during the early stage of the simulation, then the performance of the multi-directional algorithm exceeds the performance of the unidirectional.

KroAB100. As for the test cases above, for *KroAB100* the true global Pareto front was unknown. In order to obtain a global Pareto front all the runs (40) of the multi-directional D&B and unidirectional D *Physarum* algorithms were used. For this test case, only 20 runs were performed for each algorithm instead of 200.

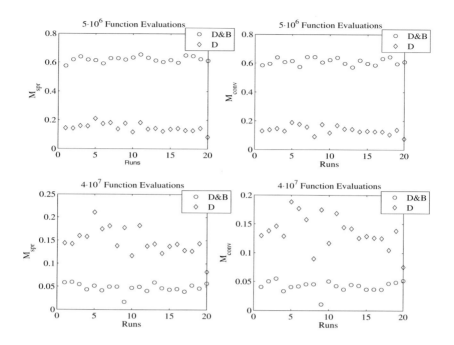

Fig. 8. Value of the metrics in Eqs. (11)-(12) for the D and D&B algorithms at $5 \cdot 10^6$ and $4 \cdot 10^7$ function evaluations

However, this number of runs doubles the one used to compare various algorithms on the same instance in [13]. In the following, *Global* is used to indicate the Pareto front found by all the runs of the multi-directional and unidirectional *Physarum* algorithms as explained above, while *Global D* and *Global D&B* are used to indicate the global Pareto fronts obtained from all the runs of the D and D&B algorithms respectively (20). Fig. 6 a) shows two examples of Pareto fronts found after one single run of the D and D&B algorithms. Fig. 6 b) shows a comparison of the *Global D* and *Global D&B*. Both figures let the reader see that the introduction of multi-directionality in the algorithm is an optimal choice. This is confirmed analyzing Fig. 7: while the indexes of performance p_{spr} and p_{conv} of the multi-directional algorithm reach respectively 88% and 75%, the ones of the unidirectional algorithm are still under 1%. However there is a small transient at a low number of function evaluations (less than $1 \cdot 10^7$), not visible in Fig. 7, where the unidirectional algorithm performs better than the multi-directional (although the performance is still not significant). This transient can be appreciated in Fig. 8: it reports the value of the metrics in Eqs. (11)-(12) for the D and D&B algorithms at $5 \cdot 10^6$ and $4 \cdot 10^7$ function evaluations. From the figure it is evident that D performs better at low function evaluations, but their increment is not able to improve its performance significantly, as for D&B. This behaviour is similar to that found in [14].

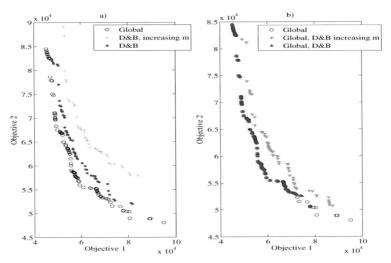

Fig. 9. Effect of the variation of the linear growth coefficient m and evaporation ρ at $4 \cdot 10^7$ function evaluations. In a) *Global* is the global Pareto front obtained from all the runs of the D and D&B algorithms (40) as in Fig. 4, *D&B* indicates an example of a Pareto front found during a run of the D&B algorithm with classical settings, as in Tab. 2, and *D&B increasing m* indicates an example of a Pareto front found during a run of the D&B algorithm where $m = 5 \cdot 10^{-3}$ and $\rho = 10^{-5}$. In b), *Global* is the global Pareto front as above, while *Global D&B increasing m* indicate the global Pareto fronts obtained from all the runs (20) of the D&B algorithms with $m = 5 \cdot 10^{-3}$ and $\rho = 10^{-4}$ and *Global, D&B* indicate the global Pareto fronts obtained from all the runs (20) of the D&B algorithms with classical settings.

Fig. 9 shows the effect of increasing the linear growth coefficient to $m = 5 \cdot 10^{-3}$ and evaporation parameter to $\rho = 10^{-5}$. This variation should increase the rate of veins' expansion with time and limit the effect of contraction. From Fig. 9 one can argue that increasing the rate of veins' expansion is a bad choice: the algorithm tends to converge very rapidly to a set of solutions and the exploration is not efficient as with lower rate of expansion.

The results obtained by applying the Physarum algorithm to the three aforementioned test cases are quite interesting and prove the initial assumption that building decision sequences in two directions and adding a matching ability is an advantageous choice if compared with the choice of building decision sequences in only one direction in the solution of multi-objective discrete decision making problems. The two *Physarum* can evaluate each step of the decision sequence from two directions and create joint paths: this forward and backward decision making process improves the performance of the algorithm.

A comparison among the proposed algorithm and other bio-inspired ACO-style algorithms is not reported in this paper and will be the subject of the future work. Recently [13] provided an excellent comparison of the performance of multi-objective ACO algorithms and of SPEA2 and NSGA-II, applied to a benchmark of TSP problems, including KroAB100. In [13] the higher performance of the

ACO algorithms if compared to NSGA-II and SPEA2 is shown. A first visual analysis of the Pareto Front obtained by the Physarum algorithm and the best multi-objective ACO algorithms in [13] indicates that the Physarum is able to find very quickly the centre of the front while there is a difficulty in finding the tails. This can be explained by the fact that a single structure is used (see Sect. 4.1).

4.1   Conclusion

This paper proposed an innovative multi-objective multi-directional incremental modified *Physarum* solver for multi-objective discrete decision making problems. The algorithm showed the ability to solve multi-objective problems in combinatorial optimisation, i.e. symmetric traveling salesman and vehicle routing problems, that were selected as representative examples of multi-objective reversible decision making problems. Simulations on selected test cases proved that a multi-directional approach with matching ability performs better than a unidirectional one when applied to small scale multi-objective reversible discrete decision making problems. This result is in line with the results showed in [14] for single-objective discrete decision making using a multi-directional modified Physarum algorithm. The multi-directional decision making process enhances the performance of the multi-objective solver: this gain is up to 50% (based on the indexes of performance proposed in Sect. 3.1) for the VRP test case.

It should be noted that, as introduced in Sect. 2.2, the strategy of using a single structure (in this case the flux), where the index of dominance I_{ij} is the parameter from which the *Physarum* draws knowledge on the decision space, is new. It has the advantage of being very easy to implement and it can be used when the number of objectives is high, as in many real-world problems. The disadvantage is that the proposed approach tends to concentrate the virtual exploring agents in the centre of the Pareto front, excluding the tails. On the other hand, the use of multiple structures, one for each objective, was well described and studied in [13]: it has the advantage of being able to expand the tails of the Pareto front, but an increase in the number of objectives would lead to the introduction of more structures, resulting in computational cost and complexity. However, the disadvantage of using a single structure as proposed in this paper, could be overcome by adding sub-populations of agents that consider only one objective. This will be further studied in the future, although first results are encouraging.

References

1. Dorigo, M., Maniezzo, V., Colorni, A.: The Ant System: optimisation by a colony of cooperating agents. IEEE Transactions on Systems, Man, and Cybernetics-Part B 26(1), 29–41 (1996)
2. Chong, C.S., Low, M.Y.H., Sivakumar, A.I., Gay, K.L.: A bee colony optimisation algorithm to job shop scheduling. In: Proceedings of the Winter IEEE Simulation Conference, WSC 2006, pp. 1954–1961 (2006)

3. Sayadi, M.K., Ramezanian, R., Ghaffari-Nasab, N.: A discrete firefly meta-heuristic with local search for makespan minimisation in permutation flow shop scheduling problems. International Journal of Industrial Engineering Computations 1(1), 1–10 (2010)

4. Yang, X.-S., Deb, S.: Cuckoo search via Levy flights. In: World Congress on Nature & Biologically Inspired Computing, NaBIC 2009, pp. 210–214. IEEE (2009)

5. Nakagaki, T., Yamada, H., Toth, A.: Maze-Solving by an Amoeboid Organism. Nature 407, 470 (2000)

6. Tero, A., Takagi, S., Saigusa, T., Ito, K., Bebber, D.P., Fricker, M.D., Yumiki, K., Kobayashi, R., Nakagaki, T.: Rules for Biologically Inspired Adaptive Network Design. Science 439, 327 (2010)

7. Adamatzky, A., MartÃnez, G.J., Chapa-Vergara, S.V., Asomoza-Palacio, R., Stephens, C.R.: Approximating Mexican highways with slime mould. Natural Computing 10(3), 1195–1214 (2011)

8. Hickey, D.S., Noriega, L.A.: Insights into Information Processing by the Single Cell Slime Mold Physarum Polycephalum. In: UKACC Control Conference (2008)

9. Tero, A., Yumiki, K., Kobayashi, R., Saigusa, T., Nakagaki, T.: Flow-Network Adaptation in Physarum Amoebae. Theory in Biosciences 127(2), 89–94 (2008)

10. Tero, A., Kobayashi, R., Nakagaki, T.: Physarum Solver: a Biologically Inspired Method of Road-Network Navigation. Physica: A Statistical Mechanics and its Applications 363(1), 115–119 (2006)

11. Alaya, I., Solnon, C., Ghedira, K.: Ant colony optimisation for multi-objective optimisation problems. In: 19th IEEE International Conference on Tools with Artificial Intelligence, ICTAI 2007, vol. 1, pp. 450–457 (2007)

12. Lopez-Ibanez, M.: Multi-objective Ant Colony optimisation. Diploma thesis, Intellectics Group, Computer Science Department, Technische Universitat Darmstadt, Germany (2004)

13. Garcia-Martinez, C., Cordon, O., Herrera, F.: A Taxonomy and an Empirical Analysis of Multiple Objective Ant Colony optimisation Algorithms for the Bi-Criteria TSP. European Journal of Operational Research 180, 116–148 (2007)

14. Masi, L., Vasile, M.: A multi-directional Modified Physarum Solver for Optimal Discrete Decision Making. In: Proceedings of International Conference on Bio-Inspired Optimisation Methods and their Applications, BIOMA, Bohinj, Slovenia (2012)

15. Dorigo, M., Gambardella, L.M.: Ant Colonies for the Traveling Salesman Problem. BioSystems 43, 73–81 (1997)

16. Monismith Jr., D.R., Mayfield, B.E.: Slime Mold as a Model for Numerical optimisation. In: IEEE Swarm Intelligence Symposium, St. Louis MO, USA (2008)

17. TSPLIB, library of instances for Traveling Salesman and Vehicle Routing Problems, Ruprecht Karls Universitaet Heidelberg,
http://comopt.ifi.uni-heidelberg.de/software/TSPLIB95/

18. Vasile, M., Zuiani, F.: MACS: An Agent-Based Memetic Multiobjective optimisation Algorithm Applied to Space Trajectory Design. Journal of Aerospace Engineering, Institution of Mechanical Engineers, Part G (September 2011)

A Combined Pareto Differential Evolution Approach for Multi-objective Optimization

Oluwatosin Olofintoye*, Josiah Adeyemo, and Fred Otieno

Department of Civil Engineering and Surveying,
Durban University of Technology,
Durban, South Africa
geotoseen@yahoo.co.uk

Abstract. In recent years, methods of multi-objective evolutionary algorithms (MOEAs) have been developed to solve problems involving the satisfaction of multiple objectives within the limits of certain constraints, yet there still exists some uncertainty about finding a generally trustworthy method that can consistently find solutions which are really close to desired objectives in all situations. In this study, a combined Pareto multi-objective differential evolution (CPMDE) algorithm is presented. The algorithm combines methods of Pareto ranking and Pareto dominance selections to implement a novel selection scheme at each generation. The ability of CPMDE in solving unconstrained, constrained and real-world optimization problems was demonstrated. Competitive results obtained from benchmarking CPMDE suggest that it is a good alternative for solving real multi-objective optimization problems.

Keywords: Pareto, multi-objective optimization, evolutionary algorithm, differential evolution, constraints.

1 Introduction

Optimization problems are ubiquitous in engineering and the sciences. Simply put, optimization is an attempt to maximize a systems desirable properties while simultaneously minimizing its undesirable characteristics [1]. Optimization also refers to the art of finding one or more feasible solutions corresponding to extreme values of one or more objectives while satisfying specified constraints. A significant portion of research and applications in the field of optimization has focused on single objective optimization, whereas most of the natural world problems involve multiple objectives which are conflicting in nature [2]. The task of finding one or more optimum solutions in an optimization problem involving more than one objective is known as multi-objective optimization problem (MOOP) [3]. In the solution of MOOPs, the aim is to simultaneously optimize a set of conflicting objectives to obtain a group of alternative trade-off solutions called Pareto-optimal or non-inferior solutions which must be considered equivalent in

---

* Corresponding author.

O. Schütze et al. (eds.), *EVOLVE - A Bridge between Probability, Set Oriented Numerics, and Evolutionary Computation III,* Studies in Computational Intelligence 500,
DOI: 10.1007/978-3-319-01460-9_10, © Springer International Publishing Switzerland 2014

the absence of specialized information concerning the relative importance of the objectives [4,5].

Differential evolution (DE) is a stochastic direct search evolution strategy optimization method that is fairly fast and reasonably robust. Since its inception in the 90's, DE has found practical applications in the solution of scientific optimization problems [6]. Due to its reported successes, its use has been extended to other types of problem domains, including multi-objective optimization [1,7]. In recent times, several researches extending the application of DE for finding solutions in the multi-objective problem domains have been reported in the literatures [8,9,10]. For example, Fan et al., [5] presented and validated a new differential evolution method for multi-objective optimization. In their study, a new selection scheme was designed to replace the existing one to enable DE applicable to solve either single objective or multi-objective optimization problems. In their selection scheme, the trial population solution is compared with its counterpart in the current population. If the trial candidate dominates the current population member it will survive to the next generation and replace the current population vector, otherwise the current population member is retained. They suggest that if the trial solution is worse than the target solution in any one of the objectives, it should be discarded. The method was validated using three multi-objective benchmark optimization problems. Simulation results show that the approach is capable of generating an approximated Pareto-front for the selected problems. To further examine the practical applicability of the proposed method, it was used to optimize a prototype air mixer subject to two objectives. Results show that the new DE approach can handle practical multi-objective problems successfully.

A comprehensive survey of the state-of-the-art on methods of multi-objective optimization using differential evolution is provided by Mezura-Montes et al., [7]. In the survey, methods that adjust the selection scheme of traditional DE to implement new selection schemes for multi-objective optimization are broadly categorized as either methods employing Pareto-ranking or Pareto-dominance approaches. Methods of Pareto-ranking for multi-objective DE assign ranks to each solution in the combined trial and target population based on their non-domination levels. Solutions on the best non-dominated front are assigned a rank of '1'; the solutions in the next set are assigned '2' and so on. Algorithms using this method often select all solutions with the best ranks for propagation to the next generation. In Pareto-dominance method for DE, ranks are not assigned, rather, a solution that wins the domination contest at an index proceeds to the next generation [7]. In this study, a novel multi-objective evolutionary algorithm (MOEA) which incorporates DE as its base algorithm is proposed. The algorithm combines the Pareto-ranking and Pareto-dominance approaches in a unique way to implement a novel selection scheme at each generation. Hence, it is named combined Pareto multi-objective differential evolution (CPMDE). Results obtained from benchmarking CPMDE show its promises as an excellent alternative method of MOEA. The remainder of this paper is structured as follows. In Section 2 we present the CPMDE algorithm. Section 3 presents methodologies adopted for benchmarking CPMDE while Section 4 presents the results of the benchmark and evaluation of CPMDE. The paper is concluded in Section 5.

2   Combined Pareto Multi-objective Differential Evolution

At each generation of CPMDE, the combined population of trial and target solutions are checked and non-dominated solutions (i.e. solutions on the best non-dominated front - with rank '1') are marked as 'non-dominated' while others are marked 'dominated'. After generating a trial population, tournaments are played between trial solutions and their counterparts in the target population at the same index. Four scenarios emerge: 1) if the trial solution is marked 'non-dominated' and the target is marked 'dominated' then the trial vector replaces the target vector and the target vector is discarded. 2) If the trial solution is marked 'dominated' and the target is marked 'non-dominated' then the trial vector is discarded. 3) If both solutions are marked 'dominated' then we resort to the method of Pareto-dominance selection where the trial vector replaces the target vector if it dominates the target or if they are non-dominated with respect to each other. 4) If both vectors are marked 'non-dominated', then a harmonic average crowding distance measure suggested by Huang et al., [11] is employed to select the solution that will proceed to the next generation. Furthermore, the crowding tournament is delayed until all solutions marked 'non-dominated' in the first three scenarios are installed in the next generation after which non-dominated solutions at the remaining indices are sorted out one at a time.

2.1   CPMDE Algorithm

The step-by-step procedure of the proposed CPMDE can be summarized in the following algorithm:

---

1. Input the required DE parameters like number of individuals in the population (Np), mutation scale factor (F), crossover probability (Cr), maximum number of iterations/generations (gMax), number of objective functions (M), number of decision variables/parameters (d), upper and lower bounds of each variable.

2. Initialize all solution vectors randomly within the limits of the variable bounds.

3. Set the generation counter, $g = 0$.

4. Generate a trial population of size Np using DE's mutation and crossover operations (Price et al., [1]).

5. Perform a domination check on the combined trial and target population and mark all non-dominated solutions as 'non-dominated' while marking others as 'dominated'.

6. Play domination tournaments at each population index. Tournaments are played by comparing trial and target solutions at the same index.

   i. If the trial solution is marked 'non-dominated' and the target is marked 'dominated' then the trial vector replaces the target vector and the target vector is discarded.

   ii. If the trial solution is marked 'dominated' and the target is marked 'non-dominated' then the trial vector is discarded.

   iii. If both solutions are marked 'dominated', then replace the target vector with the trial vector if it is dominated by the trial vector or if they are non-dominated with respect to each other.

   iv If both vectors are marked 'non-dominated', then note down the index and proceed to the next index. When all solutions marked 'non-dominated' from steps i – iii above are installed in the next generation, then sort out all solutions noted in step iv one at a time using the harmonic average crowding distance measure [11]. The solution with a greater harmonic average distance is selected to proceed to the next generation.

7. Increase the generation counter, g, by 1. i.e. g = g+1.

8. If g < gMax, then go to step 4 above else go to step 9.

9. Remove the dominated solutions in the last generation.

10. Output the non-dominated solutions.

*Note domination checks are performed using the naive and slow method suggested by Deb [3].

2.2 Visualizing the Effect of the Combined Pareto Selection Procedures on the Difference Vector Distribution

DE is based on evolution using difference vectors; therefore the difference vector distribution affects the optimization process [1]. We illustrate the impact of the distribution of difference vectors on algorithm performance as follows: Figure 1a shows a hypothetical distribution of 12 vectors in a bi-objective optimization problem where both objectives are minimized. We assume vectors 1-6 are the target vectors while vectors A-F are the trial vectors. Figure 1b shows the ranks of the solutions. To fill the six slots in the next generation, we further assume that solutions 1, 2, 3 competes against A, B, C and solutions 4, 5, 6 competes against D, E, F respectively. Following these assumptions, algorithms based solely on Pareto ranking selection (PRS), (eg. NSGA-II) will select solutions 1-6 as parents for the next generation (Figure 2a), while the procedure of CPMDE selects solutions 1, 2, 3 because they have a rank of '1' as parents for the next generation.

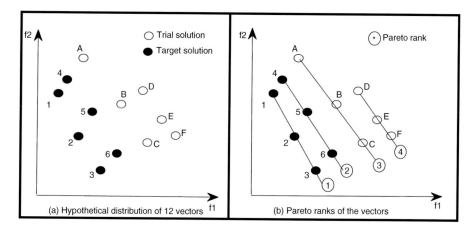

Fig. 1. Hypothetical distribution of 12 vectors and their Pareto ranks

This serves to provide a direction for the search. Also solution D will replace solution 4 while E will replace solution 5 because they are non-dominated with respect to each other though solutions 4 and 5 lies on a front with a better non-dominated rank (Figure 2b). Figures 3a and 3b present the difference vector distribution obtained by PRS and CPMDE. Inspection of Figure 3a shows that the sheaf of vector difference produced by PRS algorithms like NSGA-II contains some short vectors suitable for local search. The longer vectors are however aligned somewhat longitudinally to the best non-dominated front found. Figure 3b shows that by controlling elitism of the pool, by allowing solutions on lower

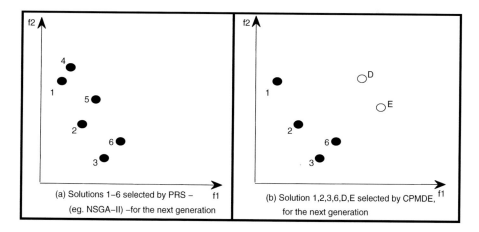

Fig. 2. Solutions selected by PRS and CPMDE for the next generation

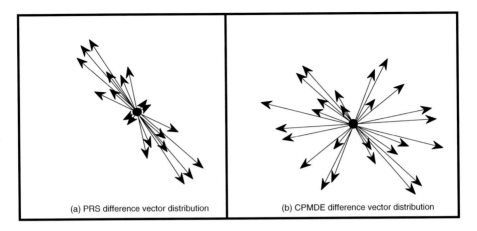

(a) PRS difference vector distribution   (b) CPMDE difference vector distribution

Fig. 3. Difference vector distributions produced by PRS and CPMDE

ranks to proceed to the next generation, the difference vector distribution of CPMDE can contain some short vectors suitable for local search and long vectors which are traverse to the fronts and suitable for a global search. Figure 4a shows that Perturbation with the difference vector distribution of procedures based solely on Pareto ranking selection like NSGA-II has a propensity to get attracted to a local optimal front while those of CPMDE are able to escape local fronts in the early generations.

2.3 Promoting Diversity among Solutions in the Obtained Non-dominated Set

In order to obtain a diverse set of solutions in the obtained non-dominated front, CPMDE employs the harmonic average crowding distance measure suggested by [11], to select the solution that will proceed to the next generation when both solutions lie on the best non-dominated front. This method harmonizes the average distances of all k-nearest neighbours around a solution. The harmonic average distance d, is computed using equation (1) [11]:

$$d = \frac{k}{\frac{1}{d_1} + \frac{1}{d_2} + \cdots + \frac{1}{d_k}} \tag{1}$$

where $d_1, d_2,, d_k$ are the Euclidean distances of k nearest neighbouring solutions and k is the number of nearest solutions. If one of the distances is very large and other distances are all small, the harmonic average distance will still be small. In this way, influence of outliers on the computation of crowding degree may be overcome. Solutions with higher harmonic average distances are better [11]. Furthermore, at higher iterations, the harmonic distance measure ensures uniform distribution of solutions on the non-dominated front.

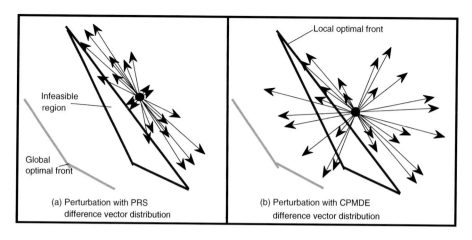

Fig. 4. Perturbation using PRS and CPMDE difference vector distributions

2.4 Handling Constraints in CPMDE

Variable Bound Constraints. In CPMDE, boundary constraints are handled using the bounce-back strategy. This strategy replaces a vector that has exceeded one or more of its bounds by a valid vector that satisfies all boundary constraints. In contrasts to random re-initialization, the bounce-back strategy takes the progress towards the optimum into account by selecting a parameter value that lies between the base vector parameter value and the bound being violated [1].

Equality and Inequality Constraints. Equality and inequality constraints are handled using the constrained-domination technique suggested by [3]. A solution $x^{(i)}$ is said to constrained-dominate a solution $x^{(j)}$, if any of the following conditions is true:

i. Solution $x^{(i)}$ is feasible and solution $x^{(j)}$ is not feasible.

ii. Solutions $x^{(i)}$ and $x^{(j)}$ are both infeasible, but solution $x^{(i)}$ has a smaller overall constraint violation.

iii. Solutions $x^{(i)}$ and $x^{(j)}$ are feasible and solution $x^{(i)}$ dominates solution $x^{(j)}$.

3   Benchmarking CPMDE

The performance of CPMDE is compared with 6 state-of-the-art MOEAs on four unconstrained benchmark test beds. The performance of CPMDE was also compared on one constrained test problem and further observed on a three-objective optimization problem. Furthermore, the performance of CPMDE on an engineering cantilever design problem is demonstrated. Other algorithms used in benchmarking CPMDE in this study include NSGA-II (real coded), NSGA-II (binary coded), SPEA, PAES, MODE-E (MODE with external archive and crowding distance measure) and MOPSO.

3.1   Benchmark Test Problems

Four test problems: SCH, FON, KUR and ZDT4 were used for evaluating the performance of CPMDE on unconstrained optimization problems. These are common difficult benchmark problems used in the literatures [10,11,12,13]. These are bi-objective problems in which both objectives are to be minimized. Each problem poses a different type of difficulty to MOEAs. SCH is a single variable problem having a convex Pareto-optimal front. This is the simplest of the test problems. FON is an n-variable problem having a non-convex Pareto-optimal front. The 3-variable version is adopted in this study. The non-convexity of the front is the major difficulty posed here. KUR is a 3-variable problem having a number of disconnected Pareto-optimal fronts. Finding uniform spread of solutions on all discontinuous regions is the challenge in this problem. ZDT4 is a 10-variable problem with 2^{19} local optimal fronts. Escaping all local non-dominated fronts to converge to the global optimal front is a real challenge in this problem.

To evaluate the performance of CPMDE on constrained optimization problem, the problem TNK is used [12]. This is a bi-objective problem with two constraints. Both of the objectives are to be minimized. TNK has a non-convex, discontinuous Pareto-optimal front. Finding uniform spread of solutions on all segments while satisfying both constraints is a challenge in this problem. Theoretical MOEA optimization studies generally consider a small number of objectives. The bi-objective case is by far the most studied. Real world MOEA applications, by contrast, are frequently more ambitious, with the number of treated criteria reaching double figures in some cases [14]. Hence, the performance of CPMDE was evaluated on test problem DTLZ2 to demonstrate its effectiveness in solving problems involving more than two objectives. The 3-objective version of the test problem is adopted in this study. The definitions and descriptions of all test functions are taken from literatures [3,8,13,14] and summarized in Table 1.

Table 1. Summary of benchmark test problems

Problem/Comments	n	Variable bounds	Objective functions and constraints	Pareto optimal solutions
SCH Convex, Unconstrained	1	$[-10^3, 10^3]$	$f_1(x) = x^2$ $f_2(x) = (x-2)^2$	$x \in [0,2]$
FON Non-convex, Unconstrained	3	$[-4,4]$	$f_1(x) = 1 - \exp\left(-\sum_{i=1}^{3}\left(x_i - \frac{1}{\sqrt{3}}\right)^2\right)$ $f_2(x) = 1 - \exp\left(-\sum_{i=1}^{3}\left(x_i + \frac{1}{\sqrt{3}}\right)^2\right)$	$x_1 = x_2 = x_3 \in \left[-\frac{1}{\sqrt{3}}, \frac{1}{\sqrt{3}}\right]$
KUR Non-convex, Discontinuous, Unconstrained	3	$[-5,5]$	$f_1(x) = \sum_{i=1}^{n-1}\left(-10\exp\left(-0.2\sqrt{x_i^2 + x_{i+1}^2}\right)\right)$ $f_2(x) = \sum_{i=1}^{n}\left(\lvert x_i\rvert^{0.8} + 5\sin x_i^3\right)$	(refer: [Deb, 2001])
ZDT4 Convex, Deceptive, Unconstrained, Multiple local-Pareto fronts	10	$x_1 \in [0,1]$ $x_i \in [-5,5]$ $i = 2,3,...,n$	$f_1(x) = x_1$ $f_2(x) = g(x)\left[1 - \sqrt{\frac{x_1}{g(x)}}\right]$ $g(x) = 1 + 10(n-1) + \sum_{i=2}^{n}\left(x_i^2 - 10\cos(4\pi x_i)\right)$	$x_1 \in [0,1]$ $x_i = 0$ $i = 2,3,...,n$
DTLZ2 Unconstrained, Three objectives	12	$[0,1]$	$f_1(x) = \left(1 + g(x)\right)\cos\left(\frac{x_1 \pi}{2}\right)\cos\left(\frac{x_2 \pi}{2}\right)$ $f_2(x) = \left(1 + g(x)\right)\cos\left(\frac{x_1 \pi}{2}\right)\sin\left(\frac{x_2 \pi}{2}\right)$ $f_3(x) = \left(1 + g(x)\right)\sin\left(\frac{x_1 \pi}{2}\right)$ $g(x) = \sum_{i=3}^{n}(x_i - 0.5)^2$	$x_1, x_2 \in [0,1]$ $x_i = 0.5$ $i = 3,4,...,n$
TNK Non-convex, Constrained, Discontinuous	2	$x_i \in [0,\pi]$ $i = 1,2$	$f_1(x) = x_1; \ \ f_2(x) = x_2$ subject to: $g_1(x) = -x_1^2 - x_2^2 + 1 + 0.1\cos\left(16 arctan\left(\frac{x_1}{x_2}\right)\right) \leq 0$ $g_2(x) = (x_1 - 0.5)^2 + (x_2 - 0.5)^2 \leq 0.5$	

3.2   Performance Measures

Various performance measures for evaluating MOEA performance have been suggested and implemented [3]. For example, Schuetze et al., [15] proposed a method for finding good Hausdorff approximations of Pareto fronts using an averaged Hausdorff distance measure (Δ_p). The measure Δ_p is a performance indicator in multi-objective evolutionary optimization which simultaneously takes into account proximity to the true Pareto front and uniform spread of solutions. Hence, it efficiently combines both spread and convergence measures in a single performance metric. The proposed methodology has further been found useful in MOEA evaluations [16,17]. In order to provide a uniform basis for comparison of MOEAs used in this study however, two performance measures reported in published studies were adopted [11,12]. Convergence metric is used to evaluate convergence to the global Pareto-optimal front while diversity metric is employed to measure the spread of solutions on the obtained non-dominated front.

Convergence Metric. This is the average distance of the non-dominated set of solutions in Q from a set P* of Pareto-optimal solutions. It is computed using equation (2). An algorithm with a smaller value of convergence metric is better [3].

$$\Upsilon = \frac{\sum_{j=1}^{|Q|} d_j}{|Q|} \tag{2}$$

where d_j is the Euclidean distance (in the objective space) between the solution $j \in Q$ and the nearest member of P*.

Diversity Metric. This metric measures the extent and spread of solutions in the obtained non-dominated front. It is computed using equation (3):

$$\Delta = \frac{\sum_{m=1}^{M} d_m^e + \sum_{i=1}^{|Q|-1} |d_i - \bar{d}|}{\sum_{m=1}^{M} d_m^e + (|Q| - 1)\bar{d}} \tag{3}$$

where d_i is the Euclidean distance (in the objective space) between consecutive solutions in the obtained non-dominated front Q, and $\bar{d}$ is the average of these distances. M is the number of objectives. The parameter d_m^e is the Euclidean distances between the extreme solution of the Pareto front P* and the boundary solution of the obtained non-dominated front Q with respect to each objective m. An algorithm with a smaller value of diversity metric Δ is better [3].

3.3   Experimental Setup

In this study, DE/rand/1/bin variant of DE was used as the base for CPMDE. Cr and F were set at 0.3. Population size Np was set to 100 and the algorithm

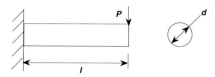

Fig. 5. A schematic diagram of a cantilever beam. *Source: Adapted from Deb, (2001)*

was run for a maximum number of generations, gMax = 250 to give a total of 25000 fitness computations. A set of 500 uniformly spaced solutions were taken from the Pareto-optimal set for computation of all metrics. Averages and variances of metric values over 10 runs are reported in this study. For test problem TNK, gMax was set at 500 generations. Harmonic average crowding distances are computed using two nearest neighbours.

3.4 Cantilever Design Problem

To demonstrate the applicability of CPMDE in solving real-world optimization problems, the algorithm was applied to design a cantilever beam. A problem originally studied by [3] using NSGA-II and further studied by [8] using MDEA is adopted here. A schematic representation of a cantilever beam is depicted in Figure 5. This problem has two decision variables of diameter(d) and length (l). The beam is designed to carry an end load P. There are two conflicting objectives that should be minimized; the weight of the beam f_1 and end deflection f_2. Minimizing the weight, f_1, will result in an optimum solution that will have small dimensions of d and l. If the dimensions are small, the beam will not be adequately rigid and the end deflection of the beam will be large. If on the other hand, the beam is minimized for end deflection, the dimensions of the beam will be large, thereby making the weight of the beam to be large. There are two constraints in this design problem. 1) The maximum stress, σ_{max} must be less than the allowable strength S_y and 2) the end deflection δ must be smaller than a specified limit of δ_{max}. The two-objective constrained optimization problem for the two decision variables d(mm) and l(mm) is formulated as follows [3]:

Objective function 1 (Minimize weight): $f_1(d, l) = \rho \frac{\pi d^2}{4} l$

Objective function 2 (Minimize deflection): $f_2(d, l) = \delta = \frac{64Pl^3}{3E\pi d^4}$

Subject to: $\sigma_{max} \leq S_y, \quad \delta \leq \delta_{max}$

Bound constraints: $10 \leq d \leq 50mm, \quad 200 \leq l \leq 1000mm$

where: $\sigma_{max} = \frac{32Pl}{\pi d^3}$

ρ, P, d and l are the density, force, diameter and length respectively. The following parameter values are used: $\rho = 7800$ kg/m^3, P $= 1$ KN, E $= 207$ GPa, Sy $= 300$ MPa and $\delta_{max} = 5$mm. On this problem, the following settings are used for CPMDE: Cr $= 0.9$, F $= 0.5$, Np $= 100$ and gMax $= 300$.

4   Results and Discussion

The mean and variance of the convergence metric on the unconstrained test beds over 10 runs of CPMDE are reported in Table 2 while those of the diversity metric are presented in Table 3. The performance metrics for MOPSO on the test problem ZDT4 is not available. The authors reported that this algorithm failed on this multi modal test bed. Reported values of convergence and diversity metrics for other algorithms used in benchmarking CPMDE are taken from correlative literatures [8,12] and presented in the respective tables. Best mean results are shown in boldface. Figure 6 depicts the convergence of the obtained non-dominated front to the true Pareto-optimal front in problems SCH and FON, while Figure 7 depicts the convergence of the obtained non-dominated front to the true Pareto-optimal front in problems KUR and ZDT4. The values of the test metrics are indicated on the respective plots. Figure 8 shows the performance of CPMDE for 500 generations on the problem TNK. Figure 9 shows the convergence of solutions obtained by CPMDE to the true Pareto-optimal surface of test problem DTLZ2 while Figure 10 shows the results obtained by CPMDE on the cantilever beam design problem.

From the results in Tables 2 and 3, it is found that CPMDE performed well in converging to the Pareto front of SCH. It produced the 3rd best result for convergence metric and the best result for diversity metric (Υ=0.003273, Δ=0.156397). PAES performed best in convergence on this test bed (Υ=0.001313), the performance of CPMDE is therefore comparable with other algorithms on this problem. CPMDE outperformed all other algorithms in converging to the Pareto fronts of test beds FON and KUR as it produced convergence metrics of Υ=0.001646 and Υ=0.017632 respectively. However, MODE-E produced better values of diversity metrics on these beds while CPMDE was the runner up in both cases. Therefore, it can be said that the performance of CPMDE is comparable to MODE-E and better than the other algorithms on these test beds.

The advantage of CPMDE in converging to the global Pareto-optimal front in deceptive multi-modal functions is amply demonstrated on test problem ZDT4. Here, CPMDE outperformed all other algorithms in convergence and diversity (Υ=0.000731, Δ=0.203378). The runner-up in this case is MODE-E with metrics (Υ=0.030689, Δ=0.338330). The convergence metric on this problem is several orders of magnitude lesser than those of other algorithms. On all unconstrained problems except KUR, CPMDE produces variance values of zero (Table 2) and a value of 0.000002 for test problem KUR. This suggests that CPMDE is reliable and stable in converging to the Pareto-optimal fronts of these beds.

Table 2. Convergence metrics on unconstrained test beds

Algorithm	Convergence metric			
	SCH	FON	KUR	ZDT4
NSGA-II (real coded)	0.003391 ± 0.000000	0.001931 ± 0.000000	0.028964 ± 0.000018	0.513053 ± 0.118460
NSGA-II (binary coded)	0.002833 ± 0.000001	0.002571 ± 0.000000	0.028951 ± 0.000016	3.227636 ± 7.307630
SPEA	0.003465 ± 0.000000	0.010611 ± 0.000005	0.049077 ± 0.000081	9.513615 ± 11.321067
PAES	$\mathbf{0.001313} \pm 0.000003$	0.151263 ± 0.000905	0.057323 ± 0.011989	0.854816 ± 0.527238
MODE-E	0.006502 ± 0.000000	0.003031 ± 0.000000	0.030819 ± 0.000008	0.030689 ± 0.004867
MOPSO	0.006603 ± 0.000000	0.002157 ± 0.000000	0.030858 ± 0.000032	N/A
CPMDE	0.003273 ± 0.000000	$\mathbf{0.001646} \pm 0.000000$	$\mathbf{0.017632} \pm 0.000002$	$\mathbf{0.000731} \pm 0.000000$

Table 3. Diversity metrics on unconstrained test beds

Algorithm	Diversity metric			
	SCH	FON	KUR	ZDT4
NSGA-II (real coded)	0.477899 ± 0.003471	0.378065 ± 0.000639	0.411477 ± 0.000992	0.702612 ± 0.064648
NSGA-II (binary coded)	0.449265 ± 0.002062	0.395131 ± 0.001314	0.442195 ± 0.001498	0.479475 ± 0.009841
SPEA	0.818346 ± 0.004497	0.804113 ± 0.002961	0.880424 ± 0.009066	0.732097 ± 0.011284
PAES	1.063288 ± 0.002868	1.162528 ± 0.008945	1.079838 ± 0.013772	0.870458 ± 0.101399
MODE-E	0.347156 ± 0.001160	$\mathbf{0.220099} \pm 0.000393$	$\mathbf{0.401911} \pm 0.000545$	0.338330 ± 0.003676
MOPSO	0.594483 ± 0.002670	0.595938 ± 0.002150	0.620227 ± 0.002170	N/A
CPMDE	$\mathbf{0.156397} \pm 0.000102$	0.308420 ± 0.000749	0.402617 ± 0.001025	$\mathbf{0.203378} \pm 0.000400$

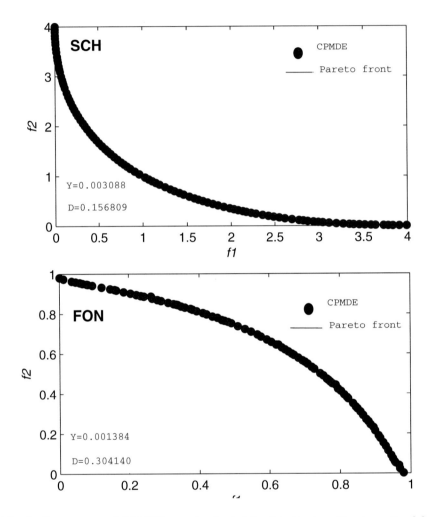

Fig. 6. Convergence of CPMDE non-dominated front to the true Pareto-optimal front in problems SCH and FON

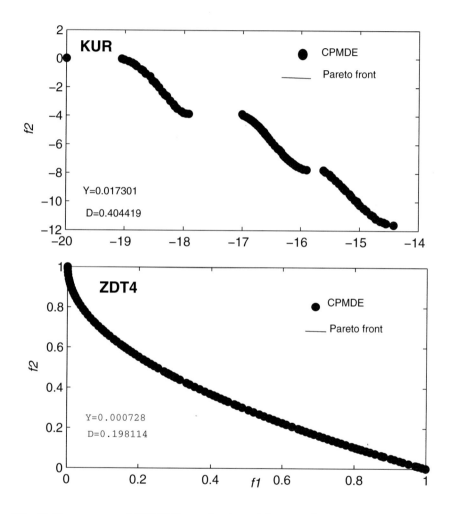

Fig. 7. Convergence of CPMDE non-dominated front to the true Pareto-optimal front in problems KUR and ZDT4

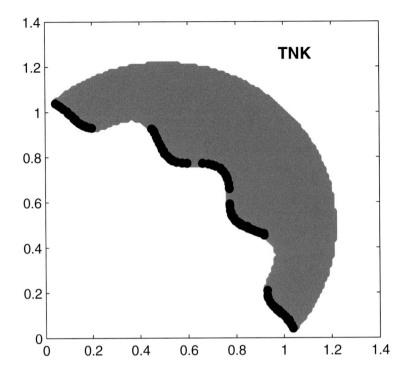

Fig. 8. Non-dominated solutions obtained by CPMDE on test problem TNK for 500 iterations

By inspection, Figure 8 shows that CPMDE is able to find uniform spread of solutions on all segments of the discontinuous Pareto front of problem TNK. This suggests that CPMDE employing a constraint domination constraint handling technique can find solutions to this type of problems.

Figure 9 depicts the convergence of the non-dominated solutions obtained by CPMDE to the true Pareto-optimal surface of DTLZ2. It can be seen from the figure that the non-dominated solutions are very close to and well distributed on the surface. Therefore, CPMDE is able to solve optimization problems involving more than two objectives.

Inspection of Figure 10 shows that CPMDE produces quality non dominated solutions along the Pareto front in an engineering cantilever design problem. This suggests that CPMDE can perform well on real-world engineering problems.

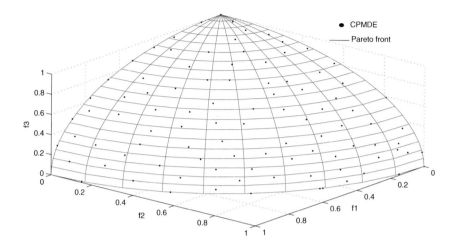

Fig. 9. Convergence of CPMDE to the true Pareto-optimal surface of DTLZ2

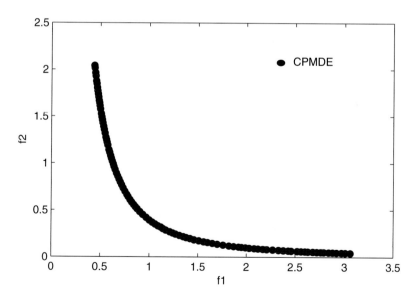

Fig. 10. Non-dominated solutions obtained by CPMDE on the cantilever design problem

5   Conclusion

In this study, a combined Pareto multi-objective differential evolution (CPMDE) multi-objective evolutionary algorithm is presented. By incorporating combined Pareto procedures to implement a novel selection scheme at each generation, CPMDE is able to adaptively balance exploitation of non-dominated solutions found with exploration of the search space. Thus it is able to escape all local optima and converge to the global Pareto-optimal front. It was found that CPMDE could converge to the Pareto optimal front of unconstrained optimization problems. The ability of CPMDE to converge to the global Pareto-optimal front in deceptive multi-modal functions is amply demonstrated on test problem ZDT4 which has 21 billion local optimal fronts. Among the 7 algorithms compared in this study, CPMDE produced the best convergence in 3 out of the 4, and best diversity in 2 out of 4 unconstrained test beds. Also, the variances of the metrics suggest that the algorithm is stable on the test beds. The ability of CPMDE in solving constrained optimization problems and optimization problems involving more than two objectives was also demonstrated. Furthermore, CPMDE was applied to solve a real-world problem where its ability to solve such problems was demonstrated. Competitive results obtained from the benchmark and application of CPMDE suggests that it is a good alternative for solving multi-objective optimization problems. Therefore, CPMDE is adoptable as a method of MOEA for solving real-world MOOPs.

References

1. Price, K.V., Storn, R.M., Lampinen, J.A.: Differential Evolution a Practical Approach to Global Optimization, 1st edn. Springer, Heidelberg (2005)
2. Babu, B.V., Chakole, P.G., Syed-Mubeen, J.H.: Multiobjective differential evolution (MODE) for optimization of adiabatic styrene reactor. Chemical Engineering Science 60, 4822–4837 (2005)
3. Deb, K.: Multi-Objective Optimization using Evolutionary Algorithms, 1st edn. John Wiley & Sons Ltd., Chichester (2001)
4. Deb, K.: Multi-Objective Optimization Using Evolutionary Algorithms: An Introduction. In: Multi-objective Evolutionary Optimisation for Product Design and Manufacturing, pp. 1–24 (2011)
5. Fan, H., Lampinen, J., Levy, Y.: An easy-to-implement differential evolution approach for multi-objective optimizations. Engineering Computations 23(2), 124–138 (2006)
6. Adeyemo, J., Olofintoye, O.: Application of hybrid models in water resources management. In: EVOLVE 2012, Mexico City, Mexico, August 07-09 (2012) ISBN 978-2-87971-112-6, ISSN 2222 - 9434
7. Mezura-Montes, E., Reyes-Sierra, M., Coello Coello, C.A.: Multi-Objective Optimization using Differential Evolution: A Survey of the State-of-the-Art. In: Chakraborty, U.K. (ed.) Advances in Differential Evolution. SCI, vol. 143, pp. 173–196. Springer, Heidelberg (2008)
8. Adeyemo, J.A., Otieno, F.A.O.: Multi-Objective Differential Evolution Algorithm for Solving Engineering Problems. Journal of Applied Sciences 9(20), 3652–3661 (2009)

9. Babu, B.V., Jehan, M.M.L.: Differential Evolution for Multi-Objective Optimization. In: Proceedings of IEEE Congress on Evolutionary Computation, Canberra, Australia, December 8-12 (2003)

10. Angira, R., Babu, B.V.: Non-dominated Sorting Differential Evolution (NSDE): An Extension of Differential Evolution for Multi-objective optimization. In: 2nd Indian International Conference on Artificial Intelligence, India (2005)

11. Huang, V.L., Suganthan, P.N., Qin, A.K., Baskar, S.: Multiobjective Differential Evolution with External Archive and Harmonic Distance-Based Diversity Measure. School of Electrical and Electronic Engineering Nanyang, Technological University Technical Report (2005)

12. Deb, K., Pratap, A., Agarwal, S., Meyarivan, T.: A Fast and Elitist Multiobjective Genetic Algorithm: NSGA-II. IEEE Transactions on Evolutionary Computation 6(2), 182–197 (2002)

13. Reddy, M.J., Kumar, D.N.: Multiobjective Differential Evolution with Application to Reservoir System Optimization. Journal of computing in Civil Engineering 21(2), 136–146 (2007)

14. Purshouse, R.C., Fleming, P.J.: Conflict, Harmony, and Independence: Relationships in Evolutionary Multi-Criterion Optimisation. In: Fonseca, C.M., Fleming, P.J., Zitzler, E., Deb, K., Thiele, L. (eds.) EMO 2003. LNCS, vol. 2632, pp. 16–30. Springer, Heidelberg (2003)

15. Schütze, O., Esquivel, X., Lara, A., Coello Coello, C.A.: Using the Averaged Hausdorff Distance as a Performance Measure in Evolutionary Multiobjective Optimization. IEEE Trans. Evolutionary Computation 16(4), 504–522 (2012)

16. Trautmann, H., Rudolph, G., Domínguez-Medina, C., Schütze, O.: Finding Evenly Spaced Pareto Fronts for Three-Objective Optimization Problems. In: Schütze, O., Coello Coello, C.A., Tantar, A.-A., Tantar, E., Bouvry, P., Del Moral, P., Legrand, P. (eds.) EVOLVE - A Bridge Between Probability, Set Oriented Numerics, and Evolutionary Computation II. AISC, vol. 175, pp. 91–106. Springer, Heidelberg (2012)

17. Rudolph, G., Trautmann, H., Schütze, O.: Homogene Approximation der Paretofront bei mehrkriteriellen Kontrollproblemen. Automatisierungstechnik 60(10), 612–621 (2012)

PSA Based Multi Objective Evolutionary Algorithms

Shaul Salomon[1], Christian Domínguez-Medina[2], Gideon Avigad[3], Alan Freitas[4], Alex Goldvard[3], Oliver Schütze[5], and Heike Trautmann[6]

[1] Department of Automatic Control and Systems Engineering, University of Sheffield, Mappin Street, Sheffield S1 3JD, UK
ssalomon1@sheffield.ac.uk

[2] Computer Research Center, CIC-IPN, Av. Juan de Dios Bátiz Esq. Miguel Othon de Mendizabal, 07738, Mexico City, Mexico
hdomigueza@sagitario.cic.ipn.mx

[3] ORT Braude College of Engineering, Snunit 51, Karmiel 21982, Israel
{gideona,goldvard}@braude.ac.il

[4] Programa de Pós-Graduação em Engenharia Elétrica, Universidade Federal de Minas Gerais, Av. Antônio Carlos 6627, 31270-901, Belo Horizonte, MG, Brasil
alandefreitas@gmail.com

[5] Computer Science Department, CINVESTAV-IPN, Av. IPN 2508, Col. San Pedro Zacatenco, Mexico City, México
schuetze@cs.cinvestav.mx

[6] Information Systems and Statistics, University of Münster, Leonardo-Campus 3, 48149 Münster, Germany
trautmann@wi.uni-muenster.de

Abstract. It has generally been acknowledged that both proximity to the Pareto front and a certain diversity along the front, should be targeted when using evolutionary multiobjective optimization. Recently, a new partitioning mechanism, the *Part and Select Algorithm* (PSA), has been introduced. It was shown that this partitioning allows for the selection of a well-diversified set out of an arbitrary given set, while maintaining low computational cost. When embedded into an evolutionary search (NSGA-II), the PSA has significantly enhanced the exploitation of diversity. In this paper, the ability of the PSA to enhance evolutionary multiobjective algorithms (EMOAs) is further investigated. Two research directions are explored here. The first one deals with the integration of the PSA within an EMOA with a novel strategy. Contrary to most EMOAs, that give a higher priority to proximity over diversity, this new strategy promotes the balance between the two. The suggested algorithm allows some dominated solutions to survive, if they contribute to diversity. It is shown that such an approach substantially reduces the risk of the algorithm to fail in finding the Pareto front. The second research direction explores the use of the PSA as an archiving selection mechanism, to improve the averaged Hausdorff distance obtained by existing EMOAs. It is shown that the integration of the PSA into NSGA-II-I and Δ_p-EMOA as an archiving mechanism leads to algorithms that are superior to base EMOAS on problems with disconnected Pareto fronts.

Keywords: multi-objective optimization, part and select algorithm (PSA), diversity, evolutionary computation.

O. Schütze et al. (eds.), *EVOLVE - A Bridge between Probability, Set Oriented Numerics, and Evolutionary Computation III,* Studies in Computational Intelligence 500,
DOI: 10.1007/978-3-319-01460-9_11, © Springer International Publishing Switzerland 2014

1 Introduction

In many real-world applications, several objectives must be optimized at the same time, leading to a multi-objective optimization problem (MOP). Mathematically, a MOP can be stated as follows:

$$\min_{\mathbf{x} \in Q} \mathbf{F}(\mathbf{x}) \qquad (1)$$

where $Q \subset \mathbb{R}^d$ is a domain in d-dimensional real space, $\mathbf{F}(\mathbf{x})$ is defined as the vector of the k objective functions:

$$\mathbf{F}(\mathbf{x}) = [f_1(\mathbf{x}), \dots, f_k(\mathbf{x})]^T$$

where each objective function $f_i(\mathbf{x}), i = 1, \dots k$, maps the vector $\mathbf{x} \in \mathbb{R}^d$ to $\mathbb{R}$. The set of optimal solutions of the problem (1) is usually called the Pareto set $\mathcal{P}$. The task of many set-oriented search procedures is to find a suitable finite sized approximation of the Pareto front $\mathbf{F}(\mathcal{P})$ (i.e., the image of the Pareto set), since this front represents the set of optimal compromises measured in objective space, which usually is of primary interest. Out of the set-oriented search procedures for the numerical treatment of MOPs, EMOAs are widely used due to their global and universal approach and their high robustness [7,11]. Most EMOAs simultaneously attempt to account for both proximity of the approximation set to the Pareto front and its diversity [4]. It has been indicated in [4] that both proximity and diversity should be explored and exploited during the evolutionary search. Exploration of diversity and proximity may be related to the selection of the next generations parents and/or the control of crossover/mutation rates [25], [1], [17]. For example, in [25] the authors suggested an adaptive variation operator that exploits the chromosomal structure (binary representation) and controls crossover/mutation rates during the evolution in order maximize the information gain and to prevent information flow disruption between the different chromosomal structures. Within the exploration phase, the authors in [5] suggested to iteratively explore for good children through iterative density estimation of different optional children combinations. In that work good candidate parents have been searched for through clustering of their related performances in the objective space. It should be noted that this procedure is applied only to non-dominated candidate solutions.

On the other hand, exploitation of proximity and diversity is related to the selection of the solutions that will be saved for the next generation (through elitism or archiving) and will take place in reproduction. Domination is the predominant approach used to exploit proximity to the true Pareto front. Diversity is exploited by different approaches that can be classified into three main categories. The first treats diversity as a property of a set and evolves sets with a good diversity. The diversity can be measured according to the accumulated distances between the members of the set [19], [27], or indirectly by the hypervolume measure [32] or the averaged Hausdorff distance Δ_p [23]. Algorithms in the second category treat diversity as a property of each individual according to the density of solutions surrounding it. Fitness sharing of NPGA [15], crowding distance of NSGA-II [9], the diversity metric based on entropy [28] and the density estimation technique used in SPEA2 [31] are examples of this category. Algorithms of the third category decompose the multi-objective problem into a number of single objective problems (scalarization). Each of these problems ideally aims for a different zone on the Pareto front such that the set of solutions to the auxiliary problems form a diverse set of optimal solutions. MOEA/D [29] is probably the most famous method within this category. A recent method from this category [33] combines Pareto dominance with Chebyshev decomposition for the selection process.

When selection takes place for the sake of exploiting proximity and diversity, proper selection criteria must be formulated, in order to achieve a balance among these two inspirations. Such a balance is not easy to achieve because it has been shown that these motivations are contradicting [3]. An improvement in one usually involves regression in the other. A balance between proximity and diversity within the exploitation phase has been targeted in various ways. One way is to select the elite population by pure truncation selection. In truncation selection, the algorithm sorts all individuals based on their domination level and includes the first individuals as the elite population. Truncation selection is exploited in many EMOAs, such as NSGA-II [9] and SPEA2 [31]. In those algorithms the exploitation of proximity takes over that of diversity as the solutions are primarily chosen based on domination relations. Some efforts to overcome this drawback have been made e.g., using the Balanced Truncation Selection (BTS) [5] within MIDEA (Multi-objective Mixture based Iterated Density Estimation Evolutionary Algorithm). In that algorithm, the exploitation of diversity can be improved by a tuned truncation threshold. The idea is to include in the elite population more diverse solutions by allowing higher truncation threshold values at the beginning of the search. It is noted that also in this algorithm, the non-dominated solutions will be preferred over dominated solutions. In other words, a solution dominated by most of the population will not be selected even though it is most isolated.

Another way to allow for a better balance between proximity and diversity is to change the dominance relation among the solutions by changing the area considered as dominated by a solution. Laumanns and Ocenasek [18] proposed to use the concept of ε-dominance [20] which is a modification of the original Pareto dominance. The underlying principle of ε-dominance is that two solutions are not allowed to be non-dominated to each other, if the difference between them is less than a properly chosen value. Extensions based on this idea are the CDAS [22], where the user can control the size of a solution's dominated area and the cone ε-dominance [3], where the shape of the dominated area is a cone.

Recently, the Part and Select Algorithm (PSA) was introduced to select a diverse subset from a given set of points [21]. This mechanism has a low computational complexity, and it is capable to select a diverse subset, of any size, even if the original set is poorly distributed. These properties make PSA suitable as a selection mechanism within EMOAs. It has been shown in [21] that the integration of the PSA into NSGA-II improves its ability to find a diverse approximated set. In [2] a niching mechanism based on the PSA was used to find a set of different cross sections for a topology optimization problem.

In this paper, the ability of the PSA to improve EMOAs is further investigated. Two research directions are explored here. The first one deals with embedding the PSA within a novel genetic algorithm. The algorithm adjusts the balance between proximity and diversity by allowing some dominated solutions to survive if they improve the diversity. The second one explores the use of the PSA as an archiving selection mechanism, to improve the averaged Hausdorff distance Δ_p obtained by existing EMOAs.

The remainder of this paper is organized as follows. The PSA is described in Section 2, and its previous utilization within an EA is briefly surveyed. In Section 3 a novel PSA based EMOA with an adjustable parameter to control the trade-off between proximity and diversity is introduced. The effect of this parameter is studied, and a comparison with NSGA-II-PSA is conducted in Section 4 to highlight the algorithm's advantage in dealing with a poor initial population. The implementation of PSA as an archiving mechanism integrated into NSGA-II-I and Δ_p-EMOA is presented in Section 5. The performance of these PSA based algorithms is compared with the original EMOAs. Finally, conclusions are drawn in Section 6.

2   PSA – Part and Select Algorithm

The Part and Select Algorithm (PSA) has been recently introduced in [21] as an algorithm for selecting m well-spread points from a set of n points. It has a low computational complexity ($O(nmk)$, where k is the dimensionality of the points), and can be used for many applications. The procedure has two steps: First, the set is partitioned into subsets so that similar members are grouped in the same subset. Next, a diverse subset is formed by selecting one member from each generated subset. The following description of the algorithm is borrowed from [21].

2.1   Partitioning a Set

The core of the PSA is the algorithm of partitioning a given set of points in the objective space into smaller subsets. In order to partition a set into m subsets, PSA performs $m-1$ divisions of one single set into two subsets. At each step, the set with the greatest dissimilarity among its members is divided. This is repeated until the desired stopping criterion is met. The criterion can be either a predefined number of subsets (i.e., the value of m) or a maximal dissimilarity among each of the subsets. The dissimilarity of a set A is defined by the measure $\varnothing A$ as follows:

Let $A := \{\mathbf{f}_1 = [f_{11}, \ldots, f_{1k}]^T, \ldots, \mathbf{f}_n = [f_{n1}, \ldots, f_{nk}]^T\} \subset \mathbb{R}^k$ (i.e., n objective vectors $\mathbf{f}_i = \mathbf{F}(\mathbf{x}_i)$ for vectors $\mathbf{x}_i \in Q$), and denote

$$a_j := \min_{i=1,\ldots n} f_{ij}, \quad b_j := \max_{i=1,\ldots n} f_{ij}, \quad \Delta_j := b_j - a_j, \quad j = 1, \ldots, k \qquad (2)$$

$$\varnothing A := \max_{j=1,\ldots k} \Delta_j \qquad (3)$$

In fact, $\varnothing A$ is the diameter of the set A in the Chebyshev metric. The size of $\varnothing A$ is a measure of the dissimilarity among the members of A, with a large $\varnothing A$ indicating a large dissimilarity among the members of A.

The pseudocode of PSA for a fixed value of m is shown in Algorithm 1. At every iteration the algorithm finds the subset with the largest diameter, and parts it into two subsets.

Algorithm 1. Partitioning a set A into m subsets

$A_1 \leftarrow A$

Evaluate $\varnothing A_1$ according to Eq. (3) and store $\varnothing A_1$ in an archive.

$i \leftarrow 2$

while $i < m$ **do**

    Find A_j and coordinate p_j such that $\varnothing A_j = \Delta_{p_j} = \max\limits_{l=1,\dots i-1} \varnothing A_l$

    Part A_j to subsets A_{j_1}, A_{j_2}:

    $A_{j1} \leftarrow \left\{ \mathbf{f} = \left[f_1, \dots, f_{p_j}, \dots, f_k\right]^T \in A_j, f_{p_j} \leq a_{p_j} + \varnothing A_j/2 \right\}$

    $A_{j2} \leftarrow \left\{ \mathbf{f} = \left[f_1, \dots, f_{p_j}, \dots, f_k\right]^T \in A_j, f_{p_j} > a_{p_j} + \varnothing A_j/2 \right\}$

    Evaluate $\varnothing A_{j1}$ and $\varnothing A_{j2}$ according to Eq. (3), and replace in the archive $\varnothing A_j$ and p_j with the pairs $\varnothing A_{j1}, \varnothing A_{j2}$ and p_{j1}, p_{j2} accordingly.

    $S \leftarrow \left\{ A_1, \dots, A_{j_1}, A_{j_2}, \dots, A_i \right\}$

    $i \leftarrow i+1$

Figure 1 demonstrates the steps of the algorithm and highlights the results obtained by its use. Consider the set of 24 points in the bi-objective space depicted in the top left panel of Figure 1. Suppose that the purpose is to partition this set into $m = 5$ subsets. The gray rectangle represents the region in the objective space that contains the solutions of the set. According to Eq. (3), the diameter of the given set is the length of the horizontal side of the rectangle. Therefore, the first partition is made by vertical incision (indicated by the vertical line in the middle of the rectangle). The results of this partition are depicted in the top right panel of Figure 1. The left subset in this panel has the greatest diameter (in horizontal direction). Therefore, the next partition is made on this subset by vertical incision. The results of this partition are depicted in the middle left panel of Figure 1. The other two panels of Figure 1 depict the results of the next two iterations of Algorithm 1.

Note that the results of the partitioning are different than the results of using a common grid in the original space. With a common grid, an initial interval in every dimension is divided into equal sections, resulting in the division of the hyperbox into smaller hypberboxes of equal space. Since the original set A does not necessarily 'cover' the entire space, each hyperbox in the grid might or might not contain a member of A. Hence, there is no way to predict which resulting grid will have the desired number of occupied boxes. In addition, there are certain limitations on the number of hyperboxes in the grid. For example, in a two-dimensional grid it is possible to create $m = \{1, 2, 4, 6, 9, 12, \dots\}$ boxes, while only a number of $m = n^2$, when n is a positive integer, will produce an even grid. With PSA, only the occupied space (marked as the gray rectangles in Figure 1) is considered. When a set A_i is partitioned into two subsets A_{i_1} and A_{i_2}, the space considered from now on is given only by the two hyperboxes circumscribing A_{i_1} and A_{i_2}. The rest of the space in A_i is discarded. Every partition increases the number of subsets by one, and therefore any desired number of subsets can be created.

2.2   Selection of a Representative Subset

Once the set A has been partitioned into the m subsets $A_1, \dots, A_m$, the 'most suitable' element from each subset must then be chosen in order to obtain a subset $A_{(r)}$ of A

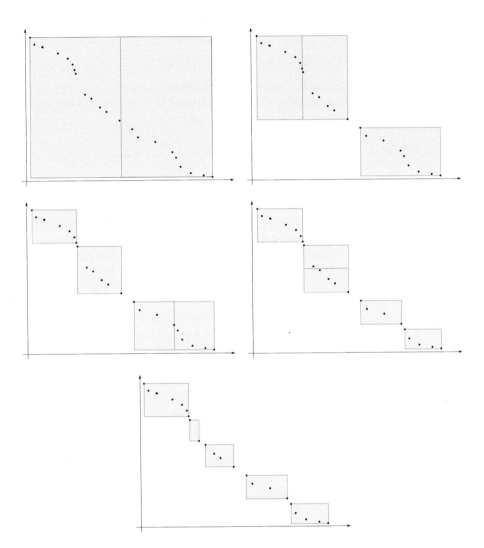

Fig. 1. Partitioning of 24 elements in bi-objective space into $m = 5$ subsets (indicated by the gray boxes) (borrowed from [21])

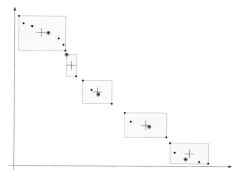

Fig. 2. Selection of a representative subset $A_{(r)}$ out of A using center point selection (borrowed from [21])

that contains m elements. This is of course problem dependent. Since this study aims for high diversity of the chosen elements, the following heuristic is suggested (denoted as center point selection): From each set A_i choose the member which is closest (in Euclidean metric) to the center of the hyperrectangle circumscribing A_i. If there exist more than one member closest to the center, one of them is chosen randomly.

Figure 2 illustrates this rule. The original set of 24 elements (compare to Figure 1) was partitioned by Algorithm 1 into five subsets. The centres of the grey rectangles are marked with a cross. In each subset the member closest to the center is circled (a random member is circled in the subset with only two members). The representative set $A_{(r)} = \{a_1, a_2, a_3, a_4, a_5\}$ is the set of all circled points.

Figure 3 illustrates the performance of PSA in selecting a subset from a randomly chosen (non-dominated) population in a three-objective space. A set of 500 randomly distributed points is depicted in Figure 3(a). The set is partitioned into 40 subsets, and the central member of each subset is selected as a representative point to form the representative subset depicted in Figure 3(f). According to Eq. (3), the diameter of the given set is the distance over f_2. Therefore, the first partition is made over f_2. At the second partition, the subset of the circles from Figure 3(b) has the largest dissimilarity and therefore is partitioned (over f_1). At the next partition the subset of gray stars is partitioned over f_1 to form the four subsets shown in Figure 3(d). The final stage of Algorithm 1 is shown in Figure 3(e). The subset shown in Figure 3(f) is obtained by selecting the point closest to the center of each of the 40 subsets. Figure 3(a) clearly shows that the distribution of the points in the original set is not uniform. Nevertheless, PSA managed to select a subset of fairly evenly distributed points from it.

2.3   NSGA-II-PSA

NSGA-II-PSA was introduced in [21] as an improvement of the well-known NSGA-II [9] by a straightforward integration of the PSA into it. The algorithm differs from its base EMOA in the selection of the elite population, and in the crowding measure assignment; both of which are conducted by using the PSA. The approximated sets obtained by NSGA-II-PSA were better then those obtained by NSGA-II in terms of both spread and convergence. Figure 4 depicts some of the comparative results between NSGA-II and NSGA-II-PSA, conducted in [21].

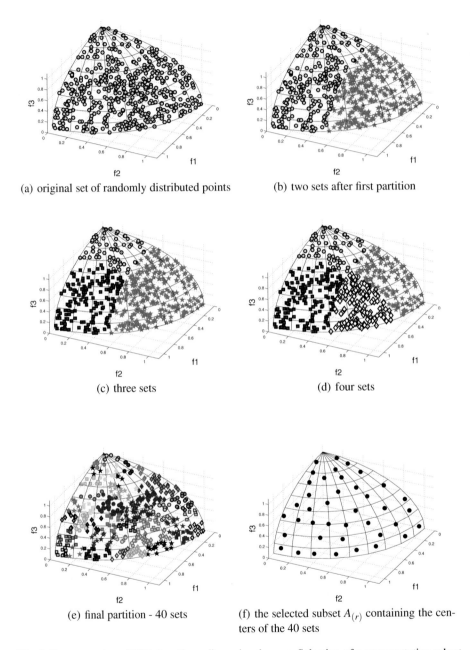

(a) original set of randomly distributed points

(b) two sets after first partition

(c) three sets

(d) four sets

(e) final partition - 40 sets

(f) the selected subset $A_{(r)}$ containing the centers of the 40 sets

Fig. 3. Demonstration of PSA in a three-dimensional space: Selection of a representative subset of 40 points from a randomly distributed set of 500 points (borrowed from [21])

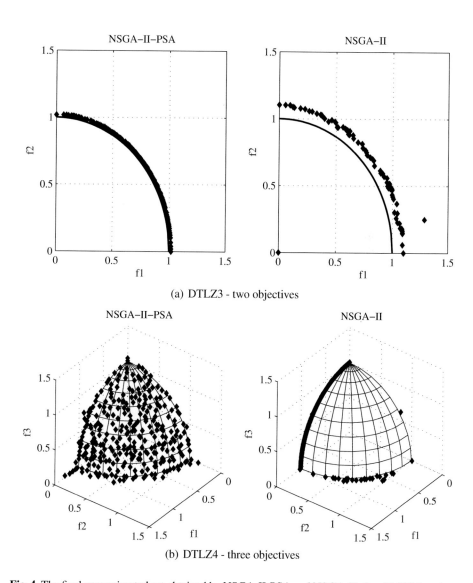

(a) DTLZ3 - two objectives

(b) DTLZ4 - three objectives

Fig. 4. The final approximated set obtained by NSGA-II-PSA and NSGA-II after 25,000 function evaluations for two objectives and 75,000 for three objectives (borrowed from [21])

3   A New EMOA with PSA as a Diversity Preservation Operator

In this section, a new EMOA is suggested - Diversity Preservation Genetic Algorithm (DPGA) - that aims simultaneously for proximity and diversity. It is designed for MOPs that pose a special challenge to spread the candidate solutions along the Pareto front. The basic structure of DPGA is similar to the structure of NSGA-II-PSA [21]. NSGA-II-PSA, as most EMOAs, inherently favors proximity over diversity. The reason for that is the selection mechanism, that selects according to non-dominance, and diversity is related as a second goal. In order to overcome this property, the selection in DPGA is conducted with two parallel goals; some solutions are selected according to their rank of non-dominance, while some solutions are selected by their remoteness from other solutions in the objective space.

In DPGA, two parent populations P_{t+1}^P and P_{t+1}^D are selected from the current population R_t. P_{t+1}^P is selected according to proximity, while P_{t+1}^D is selected according to diversity. These two populations form the new parent population: $P_{t+1} = P_{t+1}^P \cup P_{t+1}^D$. The proportion between the sizes of the two sets is controlled by the proximity factor α in the following manner: $|P_{t+1}^P| = \alpha N$, $|P_{t+1}^D| = (1-\alpha)N$, where $N = |P_{t+1}|$. The tournament selection for each population is also conducted according to its aim: Members from P_{t+1}^P are compared, as in NSGA-II-PSA, according to proximity and secondly, as a tiebreaker, according to diversity. Members from P_{t+1}^D are compared according to diversity, and secondly according to proximity. After selection, the members of both sets are combined, and crossover and mutation are applied to form the next offspring population Q_{t+1}. This procedure might produce offspring that are better both in proximity and in diversity.

By selecting according to remoteness, a highly dominated solution can be graded with a high fitness. This approach is not intuitive, and indeed, there are no methods known to the authors that give high priority to dominated solutions. Therefore, a justification of that novel approach is given here through an example. Consider the following MOP, which is a slight variation[1] of DTLZ4 for two objectives [10]:

$$
\begin{aligned}
\text{Minimize} \quad & f_1(\mathbf{x}) = r(\mathbf{x})\cos(\theta(\mathbf{x})) \\
\text{Minimize} \quad & f_2(\mathbf{x}) = r(\mathbf{x})\sin(\theta(\mathbf{x})) \\
\text{where} \quad & \mathbf{x} = [x_1,\ldots,x_7]^T \quad , \quad 0 \le x_i \le 1 \\
& \theta(\mathbf{x}) = \frac{\pi}{2}(1 - 2|x_1 - 0.5|)^{100} \\
& r(\mathbf{x}) = 1 + \sum_{i=2}^{7}(x_i - 0.5)^2
\end{aligned}
\tag{4}
$$

Proximity to the true Pareto front is defined by the value of $r(\mathbf{x})$, and the location along the Pareto front is defined by the value of $\theta(\mathbf{x})$. The Pareto optimal set corresponds

---

[1] The difference of the problem in Eq. (4) from DTLZ4 is that the peak of $\theta(x_1)$ is at $x_1 = 0.5$ rather than at $x_1 = 1$. It moves the area of interest away from the limits of the design space, which is more likely to be sampled by many EAs, including NSGA-II-PSA and DPGA.

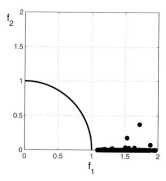

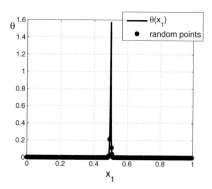

(a) Objective values of random 500 solutions. The real Pareto front is marked with a bold line.

(b) The mapping from x_1 to θ

Fig. 5. A random initial population of 500 solutions for the MOP of Eq. 4

to $r = 1$, i.e., $x_i = 0.5$ for all $i = 2, \ldots, 7$, and to all the values of θ between 0 and $\pi/2$. The mapping from x_1 to θ, as depicted in Figure 5(b), results in θ values close to zero for 98% of the x_1 values. All other values of θ correspond to $0.49 < x_1 < 0.51$. Figure 5 depicts a random population of 500 solutions. Only two solutions of this population have θ value greater than 0.1 radian. Both of them are dominated by most of the other solutions. An algorithm that favours non-dominated solutions will skip these two solutions, and their genetic information (i.e., x_1 close to 0.5), which is important to spread the approximated set along the Pareto front, will be lost.

The algorithm of DPGA is presented in Algorithm 2. A discussion about the setting of α (Step 4) appears in Section 3.3. Steps 5–7 are explained in Section 3.1. Steps 8–9 are explained in Section 3.2.

Algorithm 2. DPGA - Diversity Preservation Genetic Algorithm

1: $R_1 \leftarrow$ Generate a random set of solutions of size $2N$
2: $t \leftarrow 1$
3: **while** Stopping criteria not met **do**
4:     Set α
5:     $P_{t+1}^P \leftarrow$ Preserve αN solutions from R_t based on non-dominance
6:     $R_t^* \leftarrow R_t \setminus P_{t+1}^P$
7:     $P_{t+1}^D \leftarrow$ Preserve $(1 - \alpha)N$ solutions from R_t^* based on diversity.
8:     $Q_{t+1}^P \leftarrow S_P(P_{t+1}^P)$
9:     $Q_{t+1}^D \leftarrow S_D(P_{t+1}^D)$
10:    $Q_{t+1}^* \leftarrow Q_{t+1}^P \cup Q_{t+1}^D$
11:    $Q_{t+1}^{**} \leftarrow CrossOver(Q_{t+1}^*)$
12:    $Q_{t+1} \leftarrow Mutation(Q_{t+1}^{**})$
13:    $R_{t+1} \leftarrow P_{t+1}^P \cup P_{t+1}^D \cup Q_{t+1}$
14:    $t \leftarrow t + 1$

Algorithm 3. Elite Preservation in DPGA

$P_{t+1}^P \leftarrow$ Preserve αN solutions from R_t according to NSGA-II-PSA

assign proximity and diversity measures to the solutions in P_{t+1}^P according to
NSGA-II-PSA.

$R_t^* \leftarrow R_t \backslash P_{t+1}^P$

Partition R_t^* with PSA to $(1-\alpha)N$ subsets $\mathcal{D} = \left\{D_1, \dots, D_{(1-\alpha)N}\right\}$

$P_{t+1}^D \leftarrow \emptyset$

for each $D_i \in \mathcal{D}$ **do**

    $D_{i,nd} =$ nondominated solutions of D_i

    $\mathbf{d}_i \leftarrow$ center point selection from $D_{i,nd}$

    Assign a diversity measure to $\mathbf{d}_i$ equal to $|D_{i,nd}|$

    $P_{t+1}^D \leftarrow \left\{P_{t+1}^D, \mathbf{d}_i\right\}$

Sort P_{t+1}^D to ranks of non-dominance, and assign a proximity measure to each member according to its rank

$P_{t+1} \leftarrow P_{t+1}^P \cup P_{t+1}^D$

3.1 Elite Preservation in DPGA

At each generation DPGA preserves N members in the elite (parent) population P_{t+1}, from the current population R_t of size $2N$. This is done in two stages: First, αN members are selected from R_t to form P_{t+1}^P according to the elite preservation procedure of

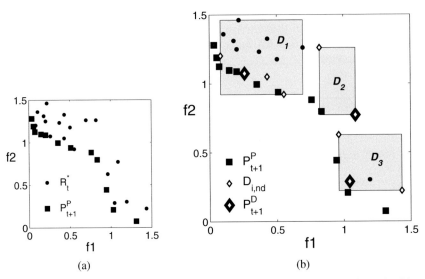

(a)   (b)

Fig. 6. Elite selection in DPGA. The 12 members preserved for proximity from the 30 members of the previous generation's population are marked with squares in (a). Preservation of 3 additional members for diversity is demonstarted in (b). The remaining members are divided into 3 sets D_1, D_2 and D_3. The non dominated members of each set $D_{i,nd}$ are marked with small diamonds. The central members of each $D_{i,nd}$, marked with large diamonds, are preserved in P_{t+1}^D

NSGA-II-PSA [21]. Next, $(1-\alpha)N$ members are selected from the remaining members in R_t to form P_{t+1}^D. This second selection is done by partitioning the remaining members of R_t to $1-\alpha$ subsets using the PSA, and including one member of each subset in P_{t+1}^D. During the elite preservation stage every member i in P_{t+1} is given a proximity measure i_{rank} and a diversity measure $i_{diversity}$. The criteria for these measures are different for the members of P_{t+1}^P and P_{t+1}^D. The exact procedure of the elite preservation and the fitness assignment is described in Algorithm 3. The procedure is illustrated in Figure 6.

3.2 Selection in DPGA

As NSGA-II, DPGA also uses a binary tournament selection from P_{t+1} to form the children population Q_{t+1}. The comparison between two candidate parents is done according to the proximity and diversity measures assigned to each member in P_{t+1}. The difference from NSGA-II is that two tournaments are done in parallel; one for the population of P_{t+1}^P, and another for P_{t+1}^D.

The diversity oriented selection operator S_D, applied on P_{t+1}^D, is described in Algorithm 4. The proximity oriented selection operator S_P, which is in fact the crowded comparison operator $\prec_n$ of NSGA-II, is the same as S_D, except for the if condition in line 4, that in the case of S_P gives the first priority to the rank of non-dominance and the second to diversity.

Algorithm 4. S_D - Diversity Oriented Selection Operator

1: $Q_{t+1}^D \leftarrow \emptyset$
2: **for** $k = 1$ to $(1-\alpha)N$ **do**
3:     Randomly select members i and j from P_{t+1}^D
4:     **if** $(i_{diversity} < j_{diversity})$ **or** $((i_{diversity} = j_{diversity})$ **and** $(i_{rank} < j_{rank}))$ **then**
5:         $Q_{t+1}^D \leftarrow Q_{t+1}^D \cup \{i\}$
6:     **else**
7:         $Q_{t+1}^D \leftarrow Q_{t+1}^D \cup \{j\}$

3.3 Sensitivity to Parameters

The performance of DPGA is highly affected by the proximity factor α. Setting α too low will hold back the algorithm from converging towards the Pareto front, since the computational power is wasted on too many dominated solutions. On the other hand, a too high value of α may lead to premature convergence, and to loss of important genetic information that may lead to undiscovered non-dominated regions. There is a stage in the evolutionary progress, when it does not make sense anymore to maintain dominated solutions, since they lack the time to reach the first front. Hence, the value of α should not be fixed for the entire run of the algorithm.

One possible way for the setting of α is suggested here. In this heuristic, DPGA consists of two stages; at the first stage a constant value of $\alpha \in (0,1)$ is set; at the second stage the selection is done as in NSGA-II-PSA (it can be conducted by simply set α to one). This heuristic requires two a-priory decisions – the value of α at the first

S. Salomon et al.

stage, and when to switch from the first to the second stage. The second decision can be described through a parameter μ – the portion of the generations in which the selection is done according to DPGA. The proper values of α and μ are problem dependent, and it is out of the authors' ability at the moment to suggest a generic way to determine them. An analysis of the performance of DPGA for one benchmark, with different values of these parameters, is given in Section 4. The conclusions on the parameters setting for this benchmark can be implemented as a starting point for other problems.

Other heuristics, such as a gradual increase of α, or setting α as a function of the generation count, can lead to better performance, but may be associated with more parameters. Probably, the proper way is to change α according to the progress of the global search. Meaning, to decrease it when the elite population loses its diversity, and to increase it otherwise. This should be done automatically within the evolutionary algorithm.

4 Simulations for DPGA

In this section, the proposed DPGA is evaluated and the sensitivity of the parameters α and μ is studied. The algorithm is analyzed on the DTLZ4 benchmark with 3 objectives. This benchmark is used, since it poses a special challenge in spreading the approximated set. This is exactly the kind of problems the DPGA should be used for. The conclusions on the parameters setting for this problem can be implemented as a starting point for other problems. The approximated sets are evaluated by the hypervolume measure (HV) [32].

First, the algorithm is tested for different values of α and μ. The values of $\alpha = \{0, 0.15, 0.3, 0.45, 0.6, 0.75, 0.9, 1\}$ and $\mu = \{0, 0.1, 0.2, 0.3, 0.4, 0.5\}$ were examined for all possible combinations. Fifty independent runs were carried out for each setting. For the sake of proper comparison, all combinations of parameter setting ran on the same fifty initial populations. The parameter setting of $\alpha = 1$ or $\mu = 0$ is the NSGA-II-PSA algorithm without the modifications of DPGA. Therefore, these settings are evaluated only once on the test set, and the corresponding results are referred to as "NSGA-II-PSA".

According to the results of this analysis, another comparison is made to check the ability of DPGA to handle a poor initial population. DPGA with the best combination of α and μ, is compared with NSGA-II-PSA as a reference in this test. Each algorithm solves the problem for 100 times with the same initial population which caused in the worst performances in the previous simulations.

4.1 Experimental Setup

Both algorithms are given real-valued decision variables. They use the simulated binary crossover (SBX) operator and polynomial mutation [8], with distribution indices of $\eta_c = 20$ and $\eta_m = 20$ respectively. A crossover probability of $p_c = 1$ and a mutation probability of $p_m = 1/3$ are used. The population size is set to 300, and the number of generations to 250.

4.2   Results of DPGA with Various Parameter Settings

The HV values of the final results in all the tests varied between 7.325 and 7.435. Approximated sets with values larger than 7.4 include at least some solutions on the surface of the sphere of the Pareto front. Sets with lower HV values consist of solutions on the $f_1 - f_2$ plane and $f_1 - f_3$ plane only. Results of that kind are considered as a failure of the algorithm to spread the approximated set along the Pareto front. Figure 7 depicts a boxplot of the statistic results of the NSGA-II-PSA ($\alpha = 1$, $\mu = 0$) and one parameter setting ($\alpha = 0.15, \mu = 0.4$), as well as three approximated fronts and their HV values. The results in Figure 7(b) are considered as a failure. Those in Figure 7(c) are quite poor, and the results in Figure 7(d) are considered as good results. The boxplot of NSGA-II-PSA in Figure 7(a) shows the failures of the algorithm as outliers. The

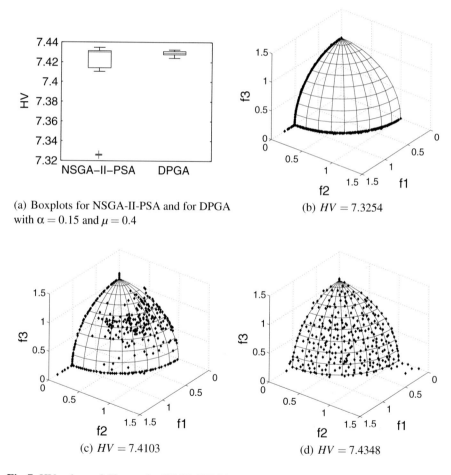

(a) Boxplots for NSGA-II-PSA and for DPGA with $\alpha = 0.15$ and $\mu = 0.4$

(b) $HV = 7.3254$

(c) $HV = 7.4103$

(d) $HV = 7.4348$

Fig. 7. HV values of 50 tests for NSGA-II-PSA and DPGA, and examples for the HV measure associated with three approximated Pareto fronts. HV values that are less than 7.4 are considered as a failure to spread the approximated set along the Pareto front (e.g., the outliers of NSGA-II-PSA, marked as red crosses, and the results in Figure 7(b)).

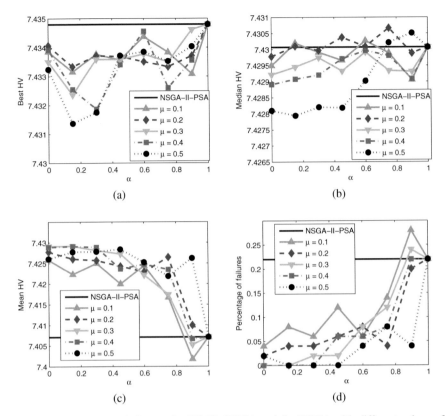

Fig. 8. Statistic results of 50 tests for NSGA-II-PSA and for DPGA with different values of α and μ

boxplots in Figure 7(a) indicate that there is no statistically significant difference in location of the HV values of NSGA-II-PSA and DPGA. On the other hand, DPGA with the above parameter setting is much more consistent regarding to different initial populations, and has no failures in spreading the approximated front. NSGA-II-PSA has 11 failures out of 50.

The results of the statistic evaluation of all the combinations of α and μ values are depicted in Figure 8. Four statistical qualities are concerned here: Figure 8(a) depicts the best HV of 50 tests; Figure 8(b) depicts the median HV; Figure 8(c) depicts the mean HV; and Figure 8(d) depicts the percentage of failures. Note that all the values of μ converge to the same point when $\alpha = 1$, since the results do not depend on μ in that case, and the algorithm is simply NSGA-II-PSA. The same statement holds for $\mu = 0$ as it is the same for all values of α (the blue line labeled "NSGA-II-PSA"). Three clear observations can be made from the results shown in Figure 8: (a) the best results from 50 trials are obtained with NSGA-II-PSA; (b) DPGA reduces the chance of a failure for most parameter settings (especially for $\alpha < 0.5$ and $\mu \geq 0.2$); (c) for this benchmark, the mean performance is more affected from the number of failures, and therefore, DPGA has a better mean HV than NSGA-II-PSA for most of the parameter settings.

These results corroborate the hypothesis that dominated solutions might contain crucial information, and the preservation of some diversified dominated solutions at the beginning of the evolutionary process can prevent premature convergence. It is worth reminding that NSGA-II-PSA is already an improvement of NSGA-II, and by recalling Figure 6(b) in [21], all of the approximated sets found with NSGA-II for DTLZ4 with 3 objectives have HV lower than 7.4.

To choose the best parameter setting of DPGA for the DTLZ4 benchmark according to these results, the main objective should be the reduction of failures. In general, the percentage of failures decreases with the increase of μ and the decrease of α. Both $\mu = 0.4$ and $\mu = 0.5$ satisfy this demand. Due to the inevitable tradeoff between proximity and diversity, the performance should be considered as well, reflected by the mean, median and best HV. Considering all the above, the best parameter setting for this benchmark is $\alpha = 0.15, \mu = 0.4$. It had no failures, and has the best performance over all the other settings with no failures.

4.3   Poor Initial Population

Here, a comparison between NSGA-II-PSA and DPGA is conducted in order to examine the ability of the algorithms to handle a very poor initial population. The initial population which produced the largest amount of failures in Section 4.2 was used as a benchmark. In this simulation, the worst initial population is given as an input to DPGA with the best combination of α and μ, and to NSGA-II-PSA, and is solved by each algorithm for 100 times.

The HV values of the obtained approximated fronts are depicted in Figure 9. The advantage of DPGA over NSGA-II-PSA is clear. While NSGA-II-PSA has failed 45 times in finding solutions on the surface of the sphere, DPGA has only failed 3 times. The HV of the successful results are quite the same for both algorithms.

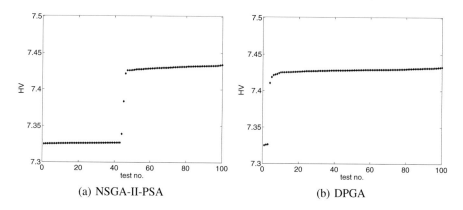

(a) NSGA-II-PSA   (b) DPGA

Fig. 9. Results of 100 tests with a poor initial population. For clarity, the results are sorted according to performance.

5   Using PSA for Hausdorff Approximations of the Pareto Front

In this section, a first attempt is made to show that PSA can be used successfully within EMOAs to compute Hausdorff approximations of the Pareto front. The Hausdorff distance d_H (e.g., [14]) prefers, roughly speaking, approximations $A \subset \mathbb{R}^n$ such that its images are located equally spaced along the Pareto front. Hence, d_H can be viewed as a performance indicator that is closely related to the terms *spread* and *convergence* as used in the EMO community. PSA is integrated into NSGAII-I [16] to produce two new algorithms: NSGA-II-I-PSA and NSGAII-I-Δ_p-PSA. Both algorithms use an external archive in addition to the procedures of NSGAII-I. In all the tests conducted in this study NSGA-II-I-PSA achieved better Hausdorff approximations[2] than its base EMOA, and NSGAII-I-Δ_p-PSA improved the performance in most cases. On models where the Pareto front is connected, both of the new methods cannot compete with Δ_p-EMOA [13], which is a specialized algorithm to produce good Hausdorff approximations. NSGA-II-I-PSA, however, is advantageous in cases where the Pareto front is disconnected. We conjecture that this is the merit of PSA that is independent from the geometry of the underlying model. First results for bi-objective problems (i.e., $k = 2$) are presented here, and considerations of $k > 2$ and further improvements of the hybrid are kept for future research.

The performance indicator considered in this section, Δ_p, is defined as follows.

Definition 1 (averaged Hausdorff distance Δ_p [23]). *Let $p \in \mathbb{N}$, $A = \{a_1,\ldots,a_r\} \subset \mathbb{R}^d$ be a candidate set and $Y = \{y_1,\ldots,y_r\} \subset \mathbb{R}^k$ be its image, i.e., $y_i = F(a_i)$, $i = 1,\ldots,r$. Further, let $P := \{p_1,\ldots,p_m\} \subset \mathbb{R}^k$ be a discretization of the Pareto front. Then it is*

$$\Delta_p(Y,P) = \max\left(\left(\frac{1}{r} \sum_{i=1}^{r} dist(y_i,P)^p \right)^{1/p}, \left(\frac{1}{m} \sum_{i=1}^{m} dist(p_i,Y)^p \right)^{1/p} \right), \quad (5)$$

where $dist(x,B) := \inf_{b\in B} \|x - b\|$ denotes the distance between a point x and a set B.

Δ_p is a combination of slight variations of the well-known *Generational Distance* (GD, see [26]) and the *Inverted Generational Distance* (IGD, see [6]). For $p = \infty$ the indicator coincides with the Hausdorff distance (i.e., $\Delta_\infty = d_H$), and hence, Δ_p can be viewed as an averaged Hausdorff distance.

The NSGA-II-I is a variant of the classical NSGA-II and is based on the conjecture that a sequential update of the crowding distances leads to a more homogeneous distribution of the population than the single determination of the crowding distances of the original NSGA-II. This algorithm is used here as a base EMOA for the new algorithms, that include an additional external archive strategy as indicated in the Figure 10.

---

[2] In fact, we will use the *averaged* Hausdorff distance in order to avoid punishments of single outliers that can occur when using stochastic search methods such as evolutionary algorithms [23].

Algorithm 5. Δ_p-Update

Require: new solution o_i, archive A_i, reference front R, archive size N_A
Ensure: new archive A_{i+1}
$\mathcal{ND}$ = nondominated solutions of $A_i \cup o_i$
if $|\mathcal{ND}| < N_A$ **then**
    for all $a \in ND$ **do**
        $h(a) = \Delta_p(ND \backslash \{a\}, R)$
        $a^* = argmin\{h(a) : a \in ND\}$
        $A_{i+1} = ND \backslash \{a^*\}$

Algorithm 6. PSA-Update

Require: population P_i, offspring O_i, archive A_i, archive size N_A
Ensure: new archive A_{i+1}
$\mathcal{ND}$ = nondominated solutions of $P_i \cup O_i \cup A_i$
if $|\mathcal{ND}| < N_A$ **then**
    $A_{i+1} = \mathcal{ND}$
else
    $A_{i+1} = PSA(\mathcal{ND}, N_A)$

Table 1. Averaged Δ_1 values for test problems with different characteristics

	Sphere model	DTLZ3	Dent	ZDT3
NSGAII-I	0.00503875	0.00638702	0.01618773	0.00591195
NSGAII-I-PSA	0.00460146	0.00621689	0.01501212	**0.00527150**
NSGAII-I-Δ_p-PSA	0.00473097	0.00680468	0.01539346	0.00552639
Δ_p-EMOA	**0.00003729**	**0.00495835**	**0.00067532**	0.00777191

PSA is being used here for the update of the archive in two variants: (i) it is used as a tool to select the best individuals to be stored in the external archive (NSGA-II-I-PSA), and (ii) PSA is integrated into the procedure of Δ_p-EMOA [13] that selects the best individuals to the external archive according to an approximated reference set (NSGAII-I-Δ_p-PSA). Here, PSA is used as a tool to obtain the reference set required to compute the distance to the set of interest. The procedure of the external archive strategy using PSA as the tool to select the best individuals in each generation (for NSGA-II-I-PSA) is detailed in Algorithm 6. The procedure where PSA is used to generate the reference set that Δ_p needs to be computed (NSGAII-I-Δ_p-PSA) is given in Algorithm 7. In this algorithm, first the set $\mathcal{ND}$ is computed that consists of all nondominated solutions of the current population P_i, the new offpsring set O_i and the current archive A_i. If the magnitude of $\mathcal{ND}$ is greater than the size of the external archive N_A, then PSA is applied on $\mathcal{ND}$ to obtain a reference front R of magnitude N_A. This set is further on used to update the archive A_i by O_i according to the best Δ_p values with respect to R. Hereby, Δ_p-Update denotes the archiver used in [13] which is given in Algorithm 5, where $h(a)$ is the Δ_p value of the set of solutions ND without the solution a.

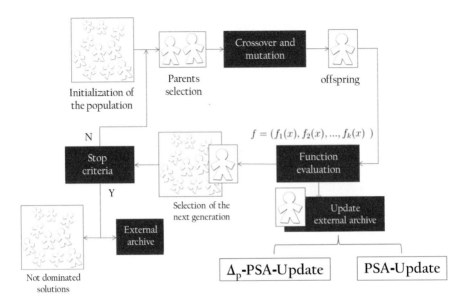

Fig. 10. General NSGAII-I procedure with an external archive. PSA-Update is being used at NSGA-II-I-PSA, and Δ_p-PSA-Update is used for NSGAII-I-Δ_p-PSA.

Algorithm 7. Δ_p-PSA-Update

Require: population P_i, offspring O_i, archive A_i, archive size N_A
Ensure: new archive A_{i+1}
  $\mathcal{ND}$ = nondominated solutions of $P_i \cup O_i \cup A_i$
  if $|\mathcal{ND}| < N_A$ **then**
    $A_{i+1} = \mathcal{ND}$
  else
    $R = \text{PSA}(\mathcal{ND}, N_A)$
    $A_{i+1} = \emptyset$
    for all $o \in O_i$ **do**
      $A_{i+1} = \Delta_p\text{-Update}(o, A_i, R)$

Table 2. Averaged Δ_1 values for test problems with disconnected fronts

	Kursawe	**Poloni**	**Schaffer**	**ZDT3**
NSGAII-I	0.03966693	0.06964843	0.02621266	0.00592310
NSGAII-I-PSA	**0.03470179**	**0.05784069**	**0.02189886**	**0.00515468**
NSGAII-I-Δ_p-PSA	0.03774589	0.06157774	0.02304279	0.00548551
Δ_p-EMOA	0.03489292	0.08614311	0.03160346	0.00778455

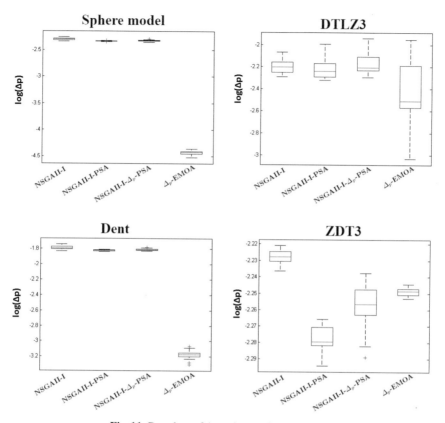

Fig. 11. Boxplots of Δ_p-values at final generation

To test the new algorithms, they are first evaluated on four test problems with different characteristics: (i) the bi-objective sphere model [13] that has a convex Pareto front, (ii) DTLZ3 [10] that has a concave Pareto front, (iii) the Dent problem [24] that has a convex-concave front, and (iv) ZDT3 [30] where the Pareto front is disconnected. The number of decision variables and their ranges are specified as recommended in literature, for the bi-objective sphere model $0 \le x_i \le 1$ ($i = 1, 2$), for the DTLZ3 $0 \le x_i \le 1$ ($i = 1, ..., 10$), for the Dent $-1.5 \le x_i \le 1.5$ ($i = 1, 2$) and for the ZDT3 $0 \le x_i \le 1$ ($i = 1, ..., 20$). Twenty independent test runs are made, each with a budget of 50,000 function calls, a population size equal to 100 and an archive size N_A equal to 100. All algorithms have been implemented in jMetal [12]. The simulated binary crossover operator is parameterized by a component-wise probability equal to 0.9 and a distribution index equal to 20. Polynomial mutation is applied using a mutation probability equal to $1/d$ (d = number of decision variables) and the distribution index equal to 20. Table 1 shows the obtained numerical results for the Δ_p indicator where $p = 1$, and Figure 11 shows boxplots of the Δ_p values at the final generation. The Δ_p indicator is calculated based on fixed reference fronts referred to as benchmark fronts in the following. Ideal benchmark fronts are composed of the set of m solutions with minimum Δ_p value with respect to the true Pareto front, where m denotes the population size of the EMOA. As the true Pareto fronts of the test problems are known in this study, these fronts are composed by m well distributed points along the true Pareto front (i.e. the set of m points with optimal PL-metric as it is defined in [13]). It can be seen that the Δ_p-EMOA yields the best results for all models with connected Pareto front, but in the case of the disconnected front (ZDT3) NSGAII-I-PSA obtains better values. This is also reflected in Figure 11 which shows the respective Δ_p-values at the final EMOA generation. Additionally, statistical significance of the results is confirmed by this means, regarding the comparison to the Δ_p-EMOA.

In order to investigate the behavior of the PSA based algorithms on models with disconnected fronts, a further test is made using the MOPs ([7]) Kursawe, Poloni, Schaffer, and ZDT3. The setting of the experiments is the same as for the previous ones. The number of decision variables and their ranges are as follows: For the Kursawe $-5 \leq x_i \leq 5$ ($i = 1,2,3$), for the Poloni $(-1 * \pi) \leq x_i \leq \pi$ ($i = 1,2$), for the Schaffer $-5 \leq x_i \leq 10$ ($i = 1$) and for the ZDT3 $0 \leq x_i \leq 1$ ($i = 1,...,20$). Table 2 shows the obtained results, and Figures 12 – 15 show the median distance to the Pareto front in terms of Δ_1 on the ordinate and the number of function evaluations on the abscissa. NSGA-II-I-PSA wins the competition on all four models which is most probably due to PSA that is independent of the geometry of the problem. Δ_p-EMOA prefers connected Pareto fronts since the reference front needed for the Δ_p archiver is built on the assumption that the Pareto front is connected [13]. Such an assumption is not made in PSA. To take into account the stochastic nature of the EMOA and to show the performance differences are significant, Figure 16 shows boxplots of the Δ_p-indicator at the final generation. The differences in location of the Δ_p-values of the NSGAII-I-PSA compared to the other EMOA are statistically significant, beside for Kursawe. These results are encouraging, however, more investigations are required to obtain a better EMOA aiming for Hausdorff approximations which we leave for future work.

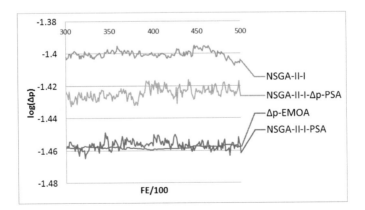

Fig. 12. Median distances to the Pareto front w.r.t. Δ_p for Kursawe problem

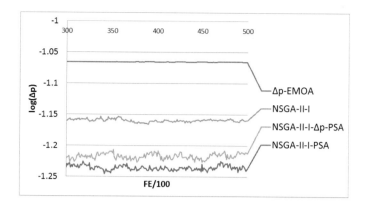

Fig. 13. Median distances to the Pareto front w.r.t. Δ_p for Poloni problem

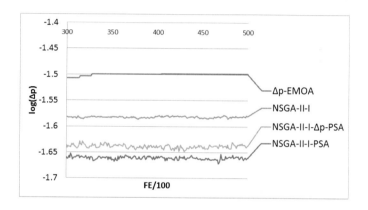

Fig. 14. Median distances to the Pareto front w.r.t. Δ_p for Schaffer problem

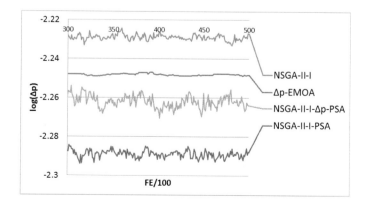

Fig. 15. Median distances to the Pareto front w.r.t. Δ_p for ZDT3 problem

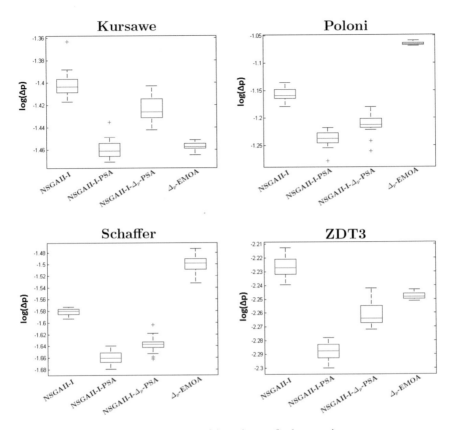

Fig. 16. Boxplots of Δ_p-values at final generation

6   Conclusions and Future Work

In this study, the ability of the PSA (Part and Select Algorithm) as a selection mechanism within EMOAs was examined. In one part of the study, PSA was used for elite selection, and in the other it was used as an archiving tool. For both cases, the results of the PSA based algorithms were satisfactory, and they were found to have better performance than their non-PSA equivalents for certain types of optimization problems.

A new evolutionary optimization approach was presented, that preserves some dominated solutions from one generation to the next. This approach was studied through a novel PSA based EMOA denoted as DPGA. The algorithm has the capacity to control the trade-off between the exploitation of proximity, to the exploitation of diversity. It was shown that by assigning high fitness to solutions that are isolated in the objective space, even if they are dominated, the chances for a failure in spreading the candidate solutions along the Pareto front decrease. As future work the performance of DPGA should be also evaluated for "regular" optimization problems that do not pose a special challenge to find a diverse set of candidate solutions. Some more comparisons with state-of-the-art EMOAs should be conducted as well. Finally, DPGA can be improved if its related parameter α could be adjusted automatically. In order to do so, a measure to identify that proximity is over-exploited on the account of diversity, is required. This measure can be used during the progress of the algorithm to decide the appropriate value of α.

The PSA was found to be also an appropriate archiving tool for Hausdorff approximations inspired EMOAs for special cases. The proposed algorithm NSGAII-I-PSA could not compete with the specialized algorithm for Hausdorff approximations Δ_p-EMOA on models where the Pareto front is connected. However in cases where the Pareto front is disconnected, NSGAII-I-PSA has outperformed the Δ_p-EMOA, producing better Hausdorff approximations to the Pareto front according to the Δ_p indicator for the four benchmark problems selected. The advantage of NSGAII-I-PSA is thanks to that PSA is independent from the geometry of the underlying problem, so the selection of the best solutions with respect to spread and convergence is not affected by the gaps within the Pareto fronts of the problems. We conjecture that the consideration of PSA will be particularly advantageous in cases more than three objectives are under consideration. Hence, the extension of the NSGA-II-I-PSA to higher-dimensional problems seems like a promising research direction.

Acknowledgements. This research was supported by a Marie Curie International Research Staff Exchange Scheme Fellowship within the 7th European Community Framework Programme.

Oliver Schütze acknowledges support from the CONACyT project no. 128554.

References

1. Abbass, H.A.: The self-adaptive Pareto differential evolution algorithm. In: Proceedings of the 2002 Congress on Evolutionary Computation, CEC 2002, vol. 1, pp. 831–836 (May 2002)
2. Avigad, G., Eisenstadt, E.M., Salomon, S., Guimar, F.G.: Evolution of contours for topology optimization. In: Schütze, O., Coello Coello, C.A., Tantar, A.-A., Tantar, E., Bouvry, P., Del Moral, P., Legrand, P. (eds.) EVOLVE - A Bridge Between Probability, Set Oriented Numerics, and Evolutionary Computation II. AISC, vol. 175, pp. 397–412. Springer, Heidelberg (2012)
3. Batista, L.S., Campelo, F., Guimarães, F.G., Ramírez, J.A.: Pareto cone ε-dominance: Improving convergence and diversity in multiobjective evolutionary algorithms. In: Takahashi, R.H.C., Deb, K., Wanner, E.F., Greco, S. (eds.) EMO 2011. LNCS, vol. 6576, pp. 76–90. Springer, Heidelberg (2011)
4. Bosman, P.A.N., Thierens, D.: The balance between proximity and diversity in multiobjective evolutionary algorithms. IEEE Transactions on Evolutionary Computation 7(2), 174–188 (2003)
5. Bosman, P.A.N., Thierens, D.: Multi-objective optimization with diversity preserving mixture-based iterated density estimation evolutionary algorithms. International Journal of Approximate Reasoning 31(3), 259–289 (2002)
6. Coello Coello, C.A., Cortés, N.C.: Solving multiobjective optimization problems using an artificial immune system. Genetic Programming and Evolvable Machines 6(2), 163–190 (2005)
7. Coello Coello, C.A., Lamont, G.B., Van Veldhuizen, D.A.: Evolutionary algorithms for solving multi-objective problems, vol. 5. Springer, Heidelberg (2007)
8. Deb, K., Agrawal, R.B.: Simulated binary crossover for continuous search space. Complex Systems 9(2), 115–148 (1995)

9. Deb, K., Pratap, A., Agarwal, S., Meyarivan, T.: A fast and elitist multiobjective genetic algorithm: NSGA-II. IEEE Transactions on Evolutionary Computation 6(2), 182–197 (2002)

10. Deb, K., Thiele, L., Laumanns, M., Zitzler, E.: Scalable multi-objective optimization test problems. In: Proceedings of the 2002 Congress on Evolutionary Computation, CEC 2002, vol. 1, pp. 825–830 (May 2002)

11. Deb, K.: Multi objective optimization using evolutionary algorithms. John Wiley and Sons (2001)

12. Durillo, J.J., Nebro, A.J.: jmetal: A java framework for multi-objective optimization. Advances in Engineering Software 42, 760–771 (2011)

13. Gerstl, K., Rudolph, G., Schütze, O., Trautmann, H.: Finding evenly spaced fronts for multi-objective control via averaging Hausdorff-measure. In: Int'l. Proc. Conference on Electrical Engineering, Computing Science and Automatic Control, CCE 2011, pp. 975–980 (2011)

14. Heinonen, J.: Lectures on Analysis on Metric Spaces. Springer, New York (2001)

15. Horn, J., Nafpliotis, N., Goldberg, D.E.: A niched Pareto genetic algorithm for multiobjective optimization. In: Proceedings of the First IEEE Conference on Evolutionary Computation, IEEE World Congress on Computational Intelligence, vol. 1, pp. 82–87 (June 1994)

16. Kukkonen, S., Deb, K.: Improved pruning of non-dominated solutions based on crowding distance for bi-objectve optimization problems. In: Proceedings of the 2006 IEEE Congress on Evolutionary Computation, pp. 1179–1186. IEEE Press (2005)

17. Laumanns, M., Rudolph, G., Schwefel, H.P.: Mutation control and convergence in evolutionary multi-object optimization. HT014601767 (2001)

18. Laumanns, M., Očenášek, J.: Bayesian optimization algorithms for multi-objective optimization. In: Guervós, J.J.M., Adamidis, P.A., Beyer, H.-G., Fernández-Villacañas, J.-L., Schwefel, H.-P. (eds.) PPSN 2002. LNCS, vol. 2439, pp. 298–307. Springer, Heidelberg (2002)

19. Li, M., Zheng, J., Xiao, G.: An efficient mufti-objective evolutionary algorithm based on minimum spanning tree. In: IEEE Congress on Evolutionary Computation, CEC 2008, IEEE World Congress on Computational Intelligence, pp. 617–624 (June 2008)

20. Loridan, P.: ε-solutions in vector minimization problems. Journal of Optimization Theory and Applications 43, 265–276 (1984)

21. Salomon, S., Avigad, G., Goldvard, A., Schütze, O.: PSA a new scalable space partition based selection algorithm for MOEAs. In: Schütze, O., Coello Coello, C.A., Tantar, A.-A., Tantar, E., Bouvry, P., Del Moral, P., Legrand, P. (eds.) EVOLVE - A Bridge Between Probability, Set Oriented Numerics, and Evolutionary Computation II. AISC, vol. 175, pp. 137–152. Springer, Heidelberg (2012)

22. Sato, H., Aguirre, H.E., Tanaka, K.: Controlling dominance area of solutions and its impact on the performance of moeas. In: Obayashi, S., Deb, K., Poloni, C., Hiroyasu, T., Murata, T. (eds.) EMO 2007. LNCS, vol. 4403, pp. 5–20. Springer, Heidelberg (2007)

23. Schütze, O., Esquivel, X., Lara, A., Coello Coello, C.A.: Using the averaged Hausdorff distance as a performance measure in evolutionary multiobjective optimization. IEEE Transactions on Evolutionary Computation 16(4), 504–522 (2012)

24. Schütze, O., Laumanns, L., Tantar, E., Coello Coello, C.A., Talbi, E.G.: Computing gap free Pareto front approximations with stochastic search algorithms. Evolutionary Computation 18(1), 65–96 (2010)

25. Tan, K.C., Chiam, S.C., Mamun, A.A., Goh, C.K.: Balancing exploration and exploitation with adaptive variation for evolutionary multi-objective optimization. European Journal of Operational Research 197(2), 701–713 (2009)

26. Van Veldhuizen, D.A.: Multiobjective Evolutionary Algorithms: Classifications, Analyses, and New Innovations. PhD thesis, Department of Electrical and Computer Engineering. Graduate School of Engineering. Air Force Institute of Technology

27. Wineberg, M., Oppacher, F.: The underlying similarity of diversity measures used in evolutionary computation. In: Cantú-Paz, E., et al. (eds.) GECCO 2003. LNCS, vol. 2724, pp. 1493–1504. Springer, Heidelberg (2003)
28. Xiaoning, S., Min, Z., Tao, L.: A multi-objective optimization evolutionary algorithm addressing diversity maintenance. In: International Joint Conference on Computational Sciences and Optimization, CSO 2009, vol. 1, pp. 524–527 (April 2009)
29. Zhang, Q., Li, H.: MOEA/D: A multiobjective evolutionary algorithm based on decomposition. IEEE Transactions on Evolutionary Computation 11(6), 712–731 (2007)
30. Zitzler, E., Deb, K., Thiele, L.: Comparison of multiobjective evolutionary algorithms: Empirical results. Evol. Comput. 8(2), 173–195 (2000)
31. Zitzler, E., Laumanns, M., Thiele, L.: SPEA2: Improving the strength Pareto evolutionary algorithm for multiobjective optimization. In: Giannakoglou, K.C., et al. (eds.) Evolutionary Methods for Design, Optimisation and Control with Application to Industrial Problems (EUROGEN 2001), pp. 95–100. International Center for Numerical Methods in Engineering, CIMNE (2002)
32. Zitzler, E.: Evolutionary Algorithms for Multiobjective Optimization: Methods and Applications. PhD thesis, Swiss Federal Institute of Technology Zurich (1999)
33. Zuiani, F., Vasile, M.: Multi agent collaborative search based on Tchebycheff decomposition. In: Computational Optimization and Applications, pp. 1–20 (March 2013)

Author Index

Printed in the United States
By Bookmasters